The Genius Groove

THE NEW SCIENCE OF CREATIVITY

by

Dr Manjir Samanta-Laughton, MD

First published by Paradigm Revolution Publishing, 2009
PO Box 113
Buxton
SK17 9WP
info@paradigmrevolution.com
www.paradigmrevolution.com

Text copyright: Manjir Samanta-Laughton 2009
Cover Design and Illustrations: James Gordon Graham
Logo design: Rob Jenkins www.firstimpression.co.uk
Author Photo: Mike Nowill Photography
www.mikenowillphotography.co.uk

ISBN 978-0-9563778-0-7

Printed in the United Kingdom by Lightning Source.

The Genius Groove

THE NEW SCIENCE OF CREATIVITY

by

Dr Manjir Samanta-Laughton, MD

Paradigm
Revolution
Publishing
EX TENEBRUM LUX

www.paradigmrevolution.com

Contents

For Charlie, Lulu and Brock

Welcome - your life just got Groovy!

Preface

I looked at my phone - six missed calls whilst I was in surgery: two local numbers.

I called the first, half recognizing it.

It was my doctor and ex-colleague.

"Manjir, I've been trying to get hold of you all day"

"I was in surgery, I had my phone switched off."

"Manjir, your father has been on the phone again. He wants me to do another mental health check. He says you've been seeing angels and behaving really strangely."

"But I saw you a couple of weeks ago. You said I was fine. Can't I come in again this week and see you?"

"I'm afraid it has gone further than that. I have a psychiatrist here. We're coming over."

An hour later, my fledgling medical career was on the brink of collapse.

Through my misery, a surge of energy and an inner knowing coursed through my being.

I stood up and said, "This will be the making of me".

Introduction

"No more running down the wrong road

Dancing to a different drum

Can't you see what's going on

Deep inside your heart?"[a]

Michael McDonald

Sweet Freedom

Would you like to enter an exciting, new magical world? We are moving into a new era for the planet - some people see it as a period of destruction, but others also see the wonderful possibilities that are growing from the chaos. It is a time of tremendous change and upheaval, where the rules of life are no longer the same - we are shifting into a new reality. We are letting go of the structures that we identified with so closely - money, jobs, cars, big houses and becoming our true authentic selves and stepping into our true power.

It is time to hear the call of your soul, time to get truly authentic with yourself. It is time to get into your Genius Groove. *The Genius Groove* is about finding our own individual creativity - why it is we exist. It is about hearing your inner voice, that desire to be something more and greater than before. Each one of us is called to fulfill a purpose. Despite the fact that our daily lives often drown it out, there is a persistent tugging at our souls: this is our passion, our purpose, our Genius Groove.

This is the life we were born to live, the work we were born to do and our urge for a life with meaning and magic. This book is about understanding that hunger, knowing the forces that might have tried to quash it, understanding how to become your true authentic self and thereby entering into a New Paradigm in your life.

This era is so unprecedented in our history that some people are feeling lost, they don't know what is happening or how to navigate the new world. There are sweeping changes moving through all areas - from our economy to religion to science. As the old systems are crumbling around us, some are looking within to finding who they truly are. By looking within, we can find our unique gifts. And when we find our Genius Groove, an amazing new life will emerge.

Many people are already entering into this new world, where there are different rules of the game. The so-called 'received wisdom' and 'that's just how things are' don't work here. There are a growing number of people who are living a new, different reality along side the old. They are finding an authentic way of living - they are in The Genius Groove.

Dancing with Myself

In my previous book, *Punk Science*, I outlined many of the scientific changes that are occurring, from the New Biology to Quantum Physics, before revealing my own theory, The Black Hole Principle - a new level of understanding of the universe.[1]

One of the responses to the book was that people wanted to know what this New Science means for our lives - how we can best utilize it. I can think of no better way to illustrate this than to describe how the New Science gives us a new view of genius and creativity, because it is our creative genius that is so fundamental to our beings, that is so globally transformative to our lives and helps us to cross the threshold into the New Paradigm.

The Genius Groove was born from my own understanding of how the universe works both from my direct mystical knowing and from my unique scientific viewpoint, backed by scientific data. Although I have discussed some of the ideas in The Genius Groove in my lectures and workshops, some have never been discussed in public or been in print before.

Since I started lecturing on the subject of the science of spirituality and, especially after the publication of my first book, quite a few people started to think that I am rather clever. Some people are saying that I am a genius.

This could not be further from my feelings about myself some years ago. In actual fact, I have spent most of my life thinking I am pretty average. Even though I worked hard in school, I never got the absolutely top accolades. I stumbled my way through medical school, barely scraping through, almost getting kicked out. I have failed numerous exams throughout my life including my driving test.

After scraping through my medical school finals, I finally became a doctor. After my house jobs/internships, I decided to train as a General Practitioner (GP) which meant doing six month posts in jobs such as Obstetrics and Gynecology and Pediatrics. During this time, numerous people told me that I would never amount to anything.

For much of my life, I just assumed I was stupid and incompetent. Yet in the last few years I have been interviewed by various television channels including the BBC and have even gone head to head with Professor Richard Dawkins on a British documentary.[2] I have published a scientific book which has gained international acclaim. In 2008, I was flown to Japan to join a top international group of just eight scientists and philosophers for a prestigious Science Symposium. Yet I haven't even got a Bachelor of Science degree, having been refused entry due to my poor performance at medical school.

How is it that a person who has been pretty average in their career all their lives suddenly has the ability to achieve such things? I simply got into my Genius Groove and having learnt many lessons along the way, I can share them with you so you understand what The Genius Groove is and how to get into your own.

Let's go all the way

This book is not another book on the power of manifestation or designed to buoy you up for five minutes only for you to come crashing down soon afterwards. We are going to be discussing a much deeper understanding of the universe, weaving in principles from the very cutting-edge of human understanding.

This book is not for the faint hearted. It may shock and disturb some people as the screens are pulled back to reveal hidden forces in society. Nor is the path of The Genius Groove an easy one - it can involve a dark night of the soul, facing hidden emotions and surrendering the ego. But the reward is a richer, more magical life and once you have stepped forward, there is no going back.

Welcome - your life just got Groovy!

Notes

Culture Club

There are many different cultures, societies and traditions on this planet. They all have their own ideas, philosophies and evolution. When I mention social situations in this book, I am generally referring to the type of culture that is dominating the world with its influence at the moment and that is what is known as 'Western Culture' with a particular focus on the general lifestyles of people in the UK. The majority of the world has some sort of tie in with this way of living, but I am well aware that it is less strong in areas e.g. people known as indigenous people. In general, I am not referring to indigenous cultures in this book when I speak of society and its rules.

Poptastic!

Some of you will notice a theme running through the book of 1980s pop music. Some people quote Lao Tzu, I'm going to be quoting Madonna and Go West. I think the reason why the music of the early 80s seemed so inspirational is because it was just that - inspirational. I have often noticed the wisdom in some of the song lyrics of that era. I have had a lot of fun incorporating this music into the book and I hope some of you are taken on a nostalgia trip too. All lyrics have been reproduced according to the Fair Use Copyright rules and are referenced at the back of the book.

Punk Science

This book is not meant to replace *Punk Science,* which contains many of the scientific references and the full explanations for the scientific assertions in this book. For a full explanation of the Black Hole Principle with more of the background evidence, please refer to that august publication!

Thank You!

So many people have helped me in my journey, they are too numerous to mention. I only hope that you know who you are and feel in your heart how you have touched mine. A big thank you to everyone that I interviewed for this book. I am so grateful to Bianca for helping me with the title, Martin and his team for doing such a good job with the logo, Mike for the photos, Gina for doing such a great job of proof reading and providing feedback as well as helping me on the path in the first place, Helen for agreeing to lend her beautiful voice to the project, Pavel and Jack for their inspiration regarding organizational theories, Mark and Jon for advising me on economics, Jane for helping with the courage to tell my story and Jenna for her companionship throughout the whole journey, sharing so many ideas as well as always being there to provide help across the pond.

I would like to give a special thanks to Jamie for his careful help and feedback on the text, his wonderful illustrations and his unending support, generosity and love.

Part One - Out of Our Brilliant Minds

Chapter 1

Groove Control

"No one knows you better

than you know yourself

Do the thing you want,

don't wait for someone else" [b]

Madonna

You are a genius! You may not think you are, but you are. Everybody is a genius in their own right, but often we feel that we are not. In this chapter we are going to be looking at the factors behind the Groove Control - why it is that we can feel out of touch with our creativity, so that we can start to get into our Genius Groove.

It's got to come from inside

Genius is not something that you have to acquire. It does not exist outside of you and you don't become one by learning other people's skills and strategies; you access it within. We often think that genius lies in other people and not us and that we have to learn their work, and even think like them to have a hope of succeeding in life.

Mimicking someone else's genius instead of discovering your own can be a fast track to disaster and low self-esteem. The very people we are encouraged to imitate would probably be horrified if you tried to be like them, because they know that they achieved success by being true to who they are. They found their own genius by being themselves and nobody can do that as well as they can.

The good news is that the same applies to you. Your genius lies in who you are: your own soul and blueprint. Nobody can be your particular type of genius. You are a genius at being you. But first you have to find out what being you is all about and what that really means. To do this, it helps to become aware of the emotional conditioning and the habits that may have developed in you over many years.

Spotlight

There is a common perception in our society that only a certain elite group of people are blessed with genius. These are the people whom a lot of us think of when we consider the word 'genius'. The word often conjures up images of a person who is like a human computer - able to perform amazing feats of mathematics in their head. Or we think of a genius as a young hotshot who makes the headlines for entering Oxford or Cambridge University, before they can legally buy a drink at the Student Union Bar. Sometimes we think of an artistic genius like Mozart, whose talent seems so unreachable for most of us.

The ultimate archetypical image of a genius in our culture seems to be Albert Einstein who has left us the legacy that a true genius must have an unruly hairstyle and mismatching socks, whilst thinking up theories of the universe that few people understand! All of these common cultural images contribute to our sense of genius as a remote and rare occurrence, bestowed on the lucky few with good genetics.

The general view is that the rest of us must make do with mediocrity as we haven't been blessed with this quirk of nature. So we view genius from the outside. Some of us may be secretly glad, watching the pressure that is often put on smart children to perform for eager parents. Mediocrity is comforting for many of us as we feel we could not handle being in the spotlight. This suits the common feeling of unworthiness so prevalent in society. So *not* being seen as a genius actually suits most of us down to the ground.

By the time we have reached adulthood most of us don't believe we are capable of anything special. However, once upon a time we all thought ourselves as the center of the universe - it is called being a child. Commonly, babies have someone pandering to their every need. We tend to be born with an initial confidence in ourselves that we can attempt anything and if we fail, it is no big deal - we just try again. If this wasn't the case then the world would be full of people shuffling around on their bottoms, having tried walking, found it too hard and therefore never attempted it again. If we go back far enough in our lives, at one point we all believed in ourselves; so what happened?

Church of the Poisoned Mind

If we examine the journey of a child to adulthood we can see something quite striking: most people we encounter throughout our lives seem to conspire in putting us into a clearly delineated box. At an early age most children enter school, where play is put aside and everything changes. Before school, children tend to lead their own world and environment through imagination, but suddenly all that is taken away in order to sit in compulsory detention every day. You stop trusting in yourself and begin listening to something external outside of you - the teacher.

Suddenly the wisdom and control in your life has shifted from an inner-directed life to an outer-directed one. You no longer trust or believe in yourself - because you are being indoctrinated not to. Pressure is put on you to find a 'career' - the box you are destined for when you leave school.

And the content of the lessons is also important in losing this self-trust. We are taught about 'clever' people throughout history who have achieved great things. They have usually been dead for long time. We are then given a cut-down version of their discoveries to regurgitate, devoid of all their original inspiration and excitement.

For example, calculus is largely considered to be invented by Sir Isaac Newton. It is taught in schools as a dry subject often devoid of context and humanity. What children in school are not often told is that Newton himself was actually a mystical person and was obsessed with the occult.[1] When put into this context, one has to wonder if his invention of calculus was actually reflecting his understanding of the deep fundamental laws of nature and wonder at the mystical realms. But this is not how this subject is taught in school.

Instead, we get the impression that other people have the ideas, other people are the geniuses and that we should just shut up and get a job, because we are not one of the lucky ones who are ever going to achieve greatness. Teaching children repeatedly about other people's achievements, whilst at the same time teaching them not to trust themselves, reinforces the sense that the locus of control is outside of the self and that other people display the quality of genius.

Compounding these factors is the genetic dogma that has played a huge part in our scientific understanding for the last hundred years: that our genes make us who we are. Until the recent sequencing of the human genome, scientists believed that all our characteristics are determined at birth by our genes and these separate us from other humans and animals. This means we cannot change the hand that we are given and, as Darwinian dogma dictates, those with the best genes will be the best at survival and the most successful, we might as well not bother trying to improve our lot.

Despite the fact that the results of human genome project have actually shown that the reality is very different, the idea of genes determining who we are still prevails, helping us to feel powerless about the hand we were dealt at birth. (More on this later.)

Religion also plays a part, with Christianity teaching that we are born as sinners and that it is only by following a set of rules that we have any hope of becoming a good person. All this conditioning is usually very successful. By the time most children have spent a few years in the school system, they have successfully learnt not to trust themselves and that someone 'other' knows best. So gradually, the young, free person that started out with their own imagination and drives, no longer trusts themselves and feels that in order to get on in life they must learn the wisdom of others to pass exams.

9 to 5

Eventually the child metamorphoses into a worker as every-thing about their education is pointed towards getting a job. A job in which yet more people will stand above them in a hierarchy and they will spend time obeying the wisdom of 'outside and other' and live in fear of doing something 'wrong' in case the said job gets lost and they will not be able to pay the bills.

Every weekend the average worker gets to glimpse a shadow of their true self before buckling down again on a Monday morning. Some have a vague feeling that they would love to have the space to really get to know who they are, but with the demands on their time, there just really isn't the chance to detox from their day-to-day living to get to grips with their inner self. Others would rather not have the time and space away from work for fear of what hidden demons might come up should they step off the busy treadmill.

Eventually the worker becomes old enough to retire. By that time, often their spark of creativity and inner direction has been crushed by a lifetime of 'outside and other'. Retirement is often filled with unfocused activities such as travel. It can feel too late for many people to remember who they are. Their true genius remains undiscovered throughout their lifetime.

Variations on this familiar story are played out repeatedly throughout the world until almost the whole human population is so tied up with the pressing issues of survival that there is no room to discover who they really are. They go throughout their life, postponing the act of getting to know themselves, until it feels too late and they no longer care - so deeply is their imagination, creativity and true genius now buried.

If we could just take a step outside this planet for a moment and take a look at the Earth as a whole, we would see that the human race is continuously caught up in the business of survival. If we could listen to the collective thoughts of humanity – how many times would we catch someone dreaming, 'one day.. one day'. Looking from this perspective, it would be easy to see how an alien who had never seen a human before would define humanity's existence as the struggle to survive.

One could even conclude that survival is all there is to being human. And for the majority of people on the planet this would effectively be true. But this is not the full truth of being a human being. Deep in your self – you know there is something more. During my time working as a medical doctor, as I sat and listened to the stories of my patients on a daily basis, I noticed the wistful look in people's eyes as they talked about life and the sense that there must be more to existence.

I heard many stories of people putting aside their own creativity to get a job and placing the task of survival first and foremost in their lives. Now you may be reading this thinking that this is just the way life is: you have to go to school, get a job, pay the mortgage etc. Any other ideas are just fantasies, but not real life.

There are actually people right now who are living a different type of reality. They may not make up the majority of the planet, but they are living beside you in a type of parallel world. They illustrate that to step out of the struggle for survival and into your true genius is possible for us all.

In order to make this transition it may help to first understand some of the factors behind how you got into this state in the first place. How did the human race become all about survival? Let's look at the situation in detail by examining some statistics from the UK quoted in 2008, before the major economic crash later that year, which means the situation may have worsened since then.

Money Changes Everything

In January 2008 The Office for National Statistics in Great Britain published figures about national salaries that were utilized in a BBC documentary program presented by father and son team, Peter and Dan Snow. [2,3] The show shocked many people because of its revelation of how much one has to earn in Great Britain to have an income in the top 10% bracket. Many people would guess at this figure to be at least £100,000 if not more. The figure is actually much lower at £46,000 – which is hardly the realm of millionaires.[4]

This means that 90% of the British population is earning less than £46,000, with 20%, equivalent to nearly 6 million people, in the low pay bracket of £10,000 or less. The average salary is £24,907 with two thirds of the population quoted as earning less than the national average and only 5,000 people in the whole country earning more than £1m.[5]

Yet Britain is one of the most expensive places in the world to live, with average house prices of around £200,000 in 2008.[6] So the gap between what most people earned and what they could afford in terms of property in 2008 was prohibitive when you consider that mortgage lenders will loan out 4 to 4.5 times your salary. This means that the two thirds of the population that earn less that the average salary will not be able to afford a house value of £112,082, let alone the average house price of £200,000.

This leaves us with an awful lot of people struggling to even put a less than average sized roof over their heads with a very negligible amount of people in Britain being able to afford a larger than average house and a more decent lifestyle.

Running up that Hill

So why is life like this? Why is life such a struggle? If you were under any illusions that it is easy to be one of the privileged few then I hope these statistics are bringing home the harsh reality of the struggle that we are in. Your answer might be that this is just how things are. This is just the way the economy is; it is in a downturn and there is nothing that we can do about it. We are led to believe that some mystical factors control our monetary system and only clever people really understand what is going on and even they cannot control or even totally predict it.

What follows may be shocking for many people, but without acknowledging these facts, you may not be able to move on. You will always stay in a state of ignorance. When you are in ignorance you cannot free your mind. It is when you are able to free your mind that you can truly get into The Genius Groove.

But the truth sounds so preposterous that many people cannot accept it. And that is part of the problem. What I am about to say does not often appear on your television sets, therefore few people believe it is true. Yet the fact that this information does not often appear in the media is exactly part of the scheme I am about to discuss.

Our monetary system is controlled alright - it is carefully controlled and manipulated by certain groups of people, namely private banking families who effectively own money and can do what they like with it. They own the media as well and make sure that this information is not easily available.

You might be thinking that, of course, this is rubbish. Governments are in control of money and their policies determine the economic down and upturns. Ultimately, money belongs to the people who democratically elect government officials who do their best to represent the country's people when in office. Are you sure this is the way it is? In the next chapter I shall give a brief overview of the way the monetary system works and how it is controlled.

Chapter 2

Spirits in a Material world

"Our so-called leaders speak

With words they try to jail you

They subjugate the meek

But it's the rhetoric of failure"[c]

The Police

The Nature of the Beast

Humanity is in the constant state of survival, not because they are not working hard enough to get out of it, but because there is a carefully orchestrated plan to keep them that way. OK, to some people that sounds extremely weird and the 'conspiracy theory' button will be hit causing some to immediately switch off. Isn't it interesting what can happen when we give something a label? We shut it in a box and don't look at it.

In fact, this reaction is also carefully orchestrated and part of the plan that I am about to discuss. Clever isn't it – to execute the plan and then make anyone feel ridiculous for actually considering it as a reality? One of the most powerful deterrents for anyone is to make them feel ridiculous for believing in something.

Have you ever come across someone who has spoken up about people controlling society? Have you noticed at some point that these people are severely ridiculed, sometimes to the point that they are ostracized from society, David Icke being a case in point. Icke is an author whose writing exposes the plan and was misinterpreted and ridiculed on a UK chat show in the early 1990s and to this day, despite his millions of fans around the world, many people in the UK will think of that incident when they hear his name.[1]

Human beings would do anything rather than risk that type of ridicule. And the people who create the plan know this, so the media, which they own, work hard to ridicule these people in public so that the rest of us never take their words seriously.

The threat of ridicule is a very effective way to protect information as people become so afraid to look into an issue in case they themselves become ridiculed. Ridicule involves being excluded from society and as human beings we have a strong need to belong to a group and not be seen as weird. This is probably a trait from our biology – as humans we survive better in groups where we can share tasks.

You see once you know a few of the tricks utilized by these people, it is actually easy to see the plan and even predict it. Why am I telling you all this? When you start to see the patterns in the plan, you can see the big picture and start to shift away from being an unconscious pawn in the game to creating a different reality for yourself. In actual fact, if you go further, as we shall see later in this book, ultimately humanity has created this situation at a collective level in order to gain a type of experience - one that is shortly going to be coming to an end.

As I said before, there is a plan by certain groups to keep you in this state of survival. Ridiculous though it might sound at first, there is some simple evidence that you can find easily which proves the point.

Our media present economics in a very mysterious way, using long words and jargon that seem very complicated. The headlines speak of us going into recessions and experts speak of the factors that are sending us there. We nod sagely, repeating the words of the experts verbatim even though we haven't got a clue as to what is really going on.

The reality is that very few people, not even senior economists, actually understand how money works. Now that sounds shocking. The fact is that there are a lot of smoke and mirrors around our money that actually stop us from understanding its true nature. But if you know where to look, it is all pretty simple, as is the reason why this information has been kept from you. And when you understand the game about money, you understand the reason why many people, including you, are kept in the business of survival which is propagated through a school system that was designed especially to create industrial workers not free, creative people.

On a deeper level it means that people go through their whole lives as drones, never breaking free to explore their own selves and their creativity. Once you understand the system you can see through it, even own it as your creation, let it go and start stepping into a new reality where your genius can be expressed.

The Money Go Round

Let's start by looking at the high street banks and lenders. You probably have a mortgage on your house, or have a loan of some sort to a financial institution like most people, right? Like most people, you probably believe that the bank has issued this loan because it has money and it is lending you some of this money at a cost that has to be paid back with interest. This is how you would expect to behave with your friends and family if you were to lend them some money and to charge some interest on it. You would not lend out money that you did not have – that would be ridiculous.

Ridiculous though that may sound, this is exactly how the banking systems work. Bank loans are created from nothing but your signature that promises to pay back the loan. [2] The money that is supposedly transferred to your bank account when you take out a loan, that you stake your house and your possessions against and work so hard to pay back with added interest, is just conjured up by the banks.[3]

Let's look at the history of how this came about. The banking system started off as being based on gold. People deposited gold with a banker and gained slips of paper in return as IOUs. These pieces of paper started to be traded and that is what has led to our paper money, although it is no longer backed with gold.[4]

What happened next was that the banker was left with deposits of gold and started to lend out the amount of deposited gold to people in the form of the more convenient paper IOU slips. He did this several times. So say he had £10 worth of gold in the bank, he would make lots of £10 loans to people using these IOU paper slips. Let's say he lent out £100 in total, so ten people were loaned £10 each.

If you can spot the problem - the banker didn't actually have £100 himself. In fact the original £10 was not his money in the first place. Yet he knew that all the people who had made deposits were rarely going to collect all the money at once, so he could lend it out to several people at once, knowing he was unlikely to be found out if everyone paid back their loans.[5]

So not only is the banker lending out money he doesn't actually have, he is charging interest on it to ten people. In effect he has created money for himself out of nowhere. Yet although his side of things was conjured up from thin air, the borrowers side of things are not; they are working to pay back the loan and if they don't, their real assets, that *do* actually have a worldly value, will be seized by the banker making him/her even richer. That means their home, car or donkey may be repossessed to pay back the loan that the banker created from nothing anyway.

The banking system today is not based on gold deposits but works very similarly. The practice of loaning out money at ten times the value that actually exists in deposits, is a system called Fractional Reserve Banking. It is usually regulated from country to country in different ways but the principle still remains - money is being created out of practically nothing.[6]

In fact our banking system today is not really based on anything tangible at all now that it is not backed by gold or silver. Look at your bank notes - I remember when they used to say, 'I owe the bearer a pound of silver' - this is not present anymore. Money exists now as just figures on a computer.

So if someone asks for a loan from their high street bank, the money can be conjured out of nowhere by and typed into the account as long as the bank has at least 10% of the loan amount deposited at a central bank. So you are working your butt off to pay back a loan that the bank has created from nothing. And then they have the cheek to charge interest on it or even take away your real assets, such as your house, if you don't pay up.

Everybody wants to rule the world

It gets worse. Not only do the individual banks behave like this and have always behaved in a similar manner, the central banks do too and they are all interlinked with each other. These central banks like the Federal Reserve Bank and the Bank of England are part of this system too. We are under the impression that the Federal Reserve Bank, for example, is part of the US government. Its own website however tells you that it is an independent body within the government.[7] It also tells you that it is a company with shareholders. It is not part of the public sector, it is a public company, but in private hands.

The Bank of England was actually the model for the banking systems used in the USA and around the world. It was created in 1694 by private investors. Even a quick internet search will reveal that it started as a private bank and although it become nationalized in the 1946, parts of its control is still with a private limited company which is exempt from having to file public accounts under specially created legislation.[8,9]

By the time of the creation of the Bank of England, there had been a long history going back and forth of wrestling the control of the monetary system from the governors of the country, which were then the royal family, to the private money lenders – the banks. Even royalty would borrow money from the banks and be beholden to them.

To try and break this control, King Henry I of England introduced the system of tally sticks. These were literally sticks of various lengths with notches cut into them to indicate their value. They caught on because King Henry decreed that they were good for paying taxes. Why did he do this? To gain control of the economy again. However, this was not to last as the bankers bought back into the system using these sticks and then established power. So the current British economy can trace itself back to a piece of wood![10]

I have mentioned tally sticks as most people have heard of them, but don't know why they were created. But their very existence represents the wrangle between the bankers and the monarchy and demonstrates how it reached such strange depths as to use pieces of wood as currency and this can be easily verified as a part of history.

So the bankers or 'Money Masters' as they are sometimes called are really in control of our political leaders even today. Only those politicians that tow the line will be supported financially. If they do not, their careers can be destroyed and there are speculations that they are disposed of in some way - through a mysterious assassination that never is fully explained, or a scandal or some other mechanism.

Just look at Gordon Brown's first action as Chancellor of the Exchequer when the Labour party came into power in 1997 in the UK. He gave the Bank of England back the control over interest rates of the nation.[11] This was barely before he had done anything else in his new job. Could this act have been a 'thank you' to the Money Masters for helping him into power? Equally politicians can be brought down at the whim of the very same people.

Throughout the ages there have been various attempts to free us from the tyranny of the Money Masters who in truth actually create and control money. But these have never lasted and now, not only are we in their grip, the media are also controlled by them so the word rarely gets out. And if it ever does, it is quickly ridiculed as a 'conspiracy theory'.

But in the current financial crisis, there are many more people, suddenly prepared to sit up and listen. In fact, the creation of this recession/depression/credit crunch is the same as all the rest. Central Banks create booms by increasing the money in circulation and people take out loans and over extend themselves. When a recession is planned, they simply contract the money supply and call in loans. This process can even hurt the high street banks, hence the recent collapse of a few. It is all quite predictable once you know what is happening and no matter how much finger pointing is done by the media and all the myriads of culprits that have been hauled up in the press, once you know it is all a smokescreen for the real puppeteers, it becomes quite simple.

Walls come tumbling down

Why am I telling you all this? Why is it relevant to finding your genius? Simply that if you understand how money works, you can start to see through the system and free your own mind. As long as you are caught in the trap that the next person who is elected to political power will be better for the economy and financial freedom is possible, then you cannot see beyond it. Finance is not governed by mysterious factors out of your control. It is quite simple really.

There are forces in this world which are keeping your life difficult. You haven't reached financial freedom because there are people making sure of it! It is not just a case of working harder and making more money. Every time people do this, up goes inflation and their money becomes worthless. This keeps people working harder and longer hours in unrewarding jobs just to stay afloat and pay for the essentials. And when you are in the state of survival you are not in your creative, expansive state. You are not in your Genius Groove.

I don't mean for you to take the information in this chapter and go form a campaign group (unless that is your particular creativity of course). I am not talking about creating rallies and political campaigns to bring the system down. This knowledge is simply an important stage in freeing your own self: becoming aware of the barriers that you have faced and the grid that you have been held in. Eventually you will be able to see the game being played out so clearly that the financial news will make you laugh out loud: it is an exquisite joke and the common person is the butt of it.

If I can recommend any type of lifestyle change to assist you, and I don't do this very often, it is to stop paying attention to the news on any type of media. You don't need it - any information you do need the universe will bring to you in some form. Watching the news is a powerful emotional trigger which keeps you held in the grip of the game instead of having a free mind. It sounds hard, but the rewards are amazing; you will get your mind back.

Set me free

You probably now understand that you have, from an early age been indoctrinated into a system that teaches you that you are bad and rotten and that you must do something to try and fix yourself – from buying consumer goods to denying yourself sex, but nothing ever seems to be enough to make you OK. By the time you leave school you are ready to take your place next to billions of others concerned with the daily business of survival. You get a sense that if only you had enough money, then you would be free, but you have not, because somehow you have not done the right thing.

Having read this last chapter some of you will realize that there is a perfectly orchestrated game to keep you captive in the system and out of touch with your creativity. Once you gain awareness and start to see through the game you can fall through the holes in the net. You can regain control of your own mind and amazing things begin to unfold. You start to find your true self and your true Genius Groove.

Ultimately, you realize that you have been the creator of everything in your life; you are not a victim. All of creation, both at a personal and collective level, is part of a loving plan to create the ultimate lessons for the expansion of our consciousness. We will be expanding on this theme of non-judgment later in this book.

But first, in the next few chapters, we will examine the nature of genius, our views about what genius is and the changing science behind our understanding of creativity.

Chapter 3

Karma Chameleon

"I'm not aware of too many things

I know what I know, if you know what I mean" [d]

Edie Brickell and the New Bohemians

I want to talk a little bit about Intelligence Quotient (IQ) and genetics, as these concepts have a stranglehold on the way in which we view intelligence and genius in the Western world.

Prime Mover

The school system and our general culture perpetuate the idea that our intelligence is based upon IQ scores. IQ is often seen to be determined at birth and there is nothing you can do about the hand you have been given. Biology adds to this sense of determinism with the doctrine that our genes decide who we are. This idea permeates many aspects of our culture. People in the general population, not just scientists, firmly believe that our characteristics are determined by our genes including our IQ: our intelligence.

A stream of news headlines tell us that scientists have made breakthroughs in finding the particular gene for a particular characteristic, with the implication that your propensity for everything from Alzheimer's disease to homosexuality is determined by your genes and therefore fixed at birth. This fits with the Darwinian ideas that we have been living with for the last hundred years - that the drives of our lives are determined by the need to perpetuate the genes of an organism and that all our actions arise for one purpose alone - to convey an evolutionary advantage.

According to this doctrine, all of our lives are focused on how to live longer and better so we can have more chances of getting mates, rearing young successfully and passing on genetics. Our entire range of behaviors, as well as the behaviors of every other species on the planet, is reduced to the needs of a molecule that exists inside us. From the point of view of orthodox biology, we are merely vehicles for the transmission of genetic material - quite a depressing and disempowering thought!

I Should Be So Lucky

According to the theory of Natural Selection, occasional random mutations occur in the genome of a species that convey an advantage over other members of the species. Random means that there is no way that you can control this or make it happen. The implication is - if you get the lucky break that makes you fitter than the others - great. If you don't - that's too bad - you must take your place amongst the mediocre of society.

Even acts of creativity are fitted into this sense of evolutionary drive. There are many theories about creative acts, which are usually described in terms of artistic endeavors, and how they convey some sort of advantage over others.[1]

This idea of genetic determinism also has huge implications for how we see intelligence and genius. We see intelligence as yet another one of those characteristics that has been determined at birth. This leads, broadly speaking, to one of the following scenarios:

1. If you happen to have been dealt a good hand, then lucky you - you are a genius! You can enter Mensa and maybe even get a Nobel prize, which in today's society seems to epitomize the highest level of human achievement.

2. If your genes and, therefore, your intelligence are not good enough, then too bad. The hand that you were dealt at birth does not make you one of the fittest of the species - you are not lucky enough to be a genius or have a high intelligence and you will never enter Mensa or get a Nobel Prize.

In summary, the current science of intelligence goes something like the following. Your genes are determined at birth, they decide on every one of characteristics including your intelligence. If you do not have a good set of genes, you are unable to change them and will go through life as an inferior person as you will never be able to improve on your intelligence as measured by an objective test called IQ.

We have been living with this story for many years and it is getting tired, as even top educationalists are pointing out.

I Can't Go For That

If we look at the beginnings of how the intelligence quotient or IQ test was invented, it becomes clear just how much it is a product of outdated ideas. According to educationalist, Sir Ken Robinson in his book, *The Element*, Alfred Binet, who was the creator of the IQ test, had a completely different intention for the test to how it is used today.[2] He was trying to identify children's special needs so that they could get the best education. In fact, Binet did not believe that intelligence is a fixed entity and dismissed such ideas as 'brutal pessimism'.

It was Lewis Terman at Stanford university who modified the Binet test in 1916 into the prototype for the modern IQ test.[3] Unfortunately, Terman was a man with highly questionable views. He believed that IQ was fixed and determined by genetics. He wanted to distinguish those whom he saw as 'feeble-minded'. In his opinion this included people of lower social classes and from certain races such as Negroes and Latinos. He believed that people with lower IQs should not be able to breed and these philosophies actually gave rise to involuntary sterilization laws in thirty American states.[4]

It is from this dodgy ground that the IQ test and all the similar philosophies that stem from it arise. We still live with the legacy of these tests that belong to another place and time and an era of science in which it was believed that people should be treated differently according to their race and their social status at birth. The idea that your intelligence is fixed at birth not only fits the legacy of the racism of a bygone age, it also fits the concepts in genetics that we have been living with for some time - that all your characteristics are determined by your genes.

But what if this isn't right? What if we found out that your characteristics including your intelligence are not determined from birth; that in fact your genes don't make you who you are. This is actually being discovered in the field of biology right now and one of the leading revolutionaries is Dr Bruce Lipton.

Upside Down

Lipton realized that the central dogma of biology - that DNA and genetic material is in control of the cell - is not true.[5] As described before, we have been living with the idea that everything there is to know about who we are, including our intelligence, is based on our genes that are determined at birth. Effectively this says that the molecules of DNA that exist inside our cells determine everything there is to know about us.

Lipton realized that actually the reverse is true. It is not the DNA inside of the cell that is in control - it is our perception of the environment. If we were to encounter a situation that is frightening, we release hormones into our bodies which get taken up by our cells at the cell boundary - the membrane.

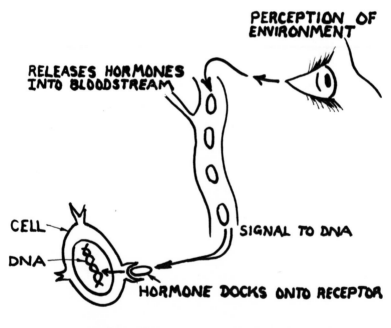

HOW PERCEPTION INFLUENCES BIOLOGY

Through a series of mechanisms, this message gets relayed to the DNA molecule in the cell. The DNA molecule may then respond to this message by altering protein production. In fact this is what DNA is all about - the production of proteins. Proteins have a myriad of functions in the body and which are determined by both the structure of the protein and how its constituent amino acids, are arranged.

DNA has the 'alphabet' and the codes to make amino acids chains in a certain order - a bit like arranging letters together to make words and then sentences - each of the individual letters matter in order for the sentence to make sense. And so it is with proteins, the amino acid building blocks gradually build up to produce longer chains which then form complex protein molecules.

These go on to perform a variety of functions according to their structure such as enzymes, receptors, neurotransmitters, structural proteins such as collagen, hormones - the list goes on and on.

It has been observed that a mistake in the DNA code leads to a change in the structure of a protein and this can lead to malfunction and disease. It is easy to see how people have become so fixated on genetics as being the source of all of who we are, because the correlations between our genetic code and diseases seem to be clear.

Too Much Monkey Business

But the concept that genes determine all that we are is seriously breaking down. The human genome project should have been one of the triumphs of orthodox biology, but actually it held many shocks for what I am calling the old scientific paradigm which we will discuss in more detail in the next chapter. The human genome project set out to outline and map all human genes and what functions they have.

It was going to be the feather in the cap of biology, but instead became a turning point, taking us into a new era of scientific understanding of ourselves. It threw up a number of serious issues that point out the flaws in the old scientific dogmas. Firstly, we found out that we don't have all that many genes. We have only about 30,000 whereas seemingly less complex creatures such as worms and fruit flies have about 15,000-20,000. [6,7] Rice plants can have 60,000 genes, twice as many as we do.[8] If genes truly make us who we are, we would expect to find that humans have 100,000 or more genes as we seem to be much more complex than worms.

The next shock was that we share over 95% of our genome with most other mammals.[9] There is very little to distinguish us from chimpanzees or even mice.[10] Where we *do* differ from chimps is in segments of DNA that code for digestive enzymes and stretches that are thought to alter brain development. One could more accurately state that the distinctions between animals are enabled by proteins that are coded by DNA, but the proteins are not the cause of the change. The differences between species are helped along by the existence of different physical molecules, but are not caused by them. The molecules simply follow suit to something deeper.

Another shock was that most of the human genome, over 90% of it, does not contain sequences that code for proteins at all.[11] These other segments are sometimes called 'junk' DNA. All this has left biologists thinking that it is not the genes that code for proteins that are important in themselves, but the way in which they are switched on and off by other segments of the genome that distinguishes between the species.

The era of genes making up who we are is coming to a close. As Lipton points out, and mainstream biologists are starting to realize, the locus of control of the cell does not lie within our DNA, so we have to start looking elsewhere.[12]

This crumbling of the central dogma of DNA has huge implications for the way in which we view intelligence. As we have seen, the concept of IQ has arisen from this sense of genetic determinism. If, due to new evidence, we realize that this concept, along with many ideas of the old mechanistic scientific paradigm, is not the true picture of the universe, then we also have to rethink our concept of intelligence and genius.

The Jean Genie

Bruce Lipton goes further than destroying the central dogma; he also says that *perception* of the environment is what causes a response in our bodies.[13] The thing is, we all perceive the same situation differently according to our underlying beliefs: some people find snakes frightening and will have a hormonal and cellular response accordingly. Some people are not troubled by snakes and have no problem being around them - they may even like them. Their bodies will have a different response to seeing a snake.

Now imagine that a child under the age of five or six is told by a well-meaning parent that snakes are dangerous and harmful and that one should never go near them. As Lipton explains, the brain waves of children are different, reflecting that they are in a different state of consciousness.[14] It is as if young children are in a permanent trance and, just like an adult who is under hypnosis, children are highly programmable. This is why they repeat words that have been said to them verbatim and sometimes in embarrassing circumstances! Whatever is said to a child is more likely to stay with them and the idea of snakes being dangerous can become a permanent belief, even if it is subconscious, that is carried throughout life, long after the original source of the belief has been forgotten.

Beliefs can therefore alter the perception of the environment which then alters the body's response to a stimulus and the way in which the cells respond! The amazing thing is, if you are able to uncover the beliefs that have gone into your system, like hidden software into a computer whilst you were a child, you can become conscious of them and therefore change your beliefs and your responses.

If you remain unaware of your beliefs, they can be triggered by situations and your body will respond in certain ways. Unless you become aware of this, you are unable to truly control it. This is why people can sometimes act out of character in certain situations. In later sections, I will go further into how emotions are an aspect of consciousness with their own dynamic and how they create these beliefs and triggers.

If you become aware of your underlying beliefs and emotions by using various methods such as psychological therapy, you can change them and therefore not be so emotionally triggered by certain situations. This new emerging view of biology, in which beliefs determine our biological responses is a revolution in science which is gradually spreading through the work of Dr Bruce Lipton and others.

This picture of biology is very different from the deterministic, mechanistic ideas that we once had. It changes the way in which we see ourselves. The locus of control of our characteristics and our emotions are not controlled at birth by a hand that we cannot change. In fact we are realizing that control of our behavior does not lie with something physical that we can dissect, but something more nebulous, such as perception and consciousness. We are moving from the idea that our biology is fixed, to the concept that our biology is fluid and ever changing.

New Moon on Monday

The old idea of fixed IQ that is determined at birth belongs to a different era - the era of mechanistic science and, dare I say it, the age of eugenics and colonialism. The legacy of a fixed IQ and the concept that we can somehow measure a person's intelligence remains with us in our schooling systems and in our wider culture.

The old eleven plus exam in the UK which used to determine a child's level of schooling remains an open wound in the people who were at the receiving end of a low mark. Robinson reports how, in his experience, he has come across people who remain convinced years later that they are not intelligent on the basis of this one exam - such is the power of these so-called objective tests.[15]

It is time to throw away the old ideas of intelligence for good. Intelligence isn't fixed at birth, nor is it the result of having been born with the best machinery that somehow conveys an advantage. It cannot be measured by objective tests that determine if you are or if you are not intelligent and there is not just one type of intelligence. These are ideas that belong to a scientific era that is fading.

In the new era of science, the evidence shows us that our fate does not lie in our genes, but that our DNA is a fluid process: with our bodies responding to our environment according to underlying beliefs that can be changed. As we shall see, our emotions are a tangible force that can be running the show so we can become aware of them and change our lives.

A molecule that was handed down to us from our parents does not define us - we are multidimensional beings in a universe that is conscious and alive! Each of us has our own unique brilliance and intelligence that cannot be measured or defined. We each have our own Genius Groove.

In the next chapter we will explore the current scientific revolution even further in order to reveal the New Science of genius and creativity.

Part Two - She Blinded me with Science

Chapter 4

Revolution Baby

"I'm so tired of all this confusion

Everyone talking like there's no solution

When a change of heart

could be the revolution baby"[e]

Transvision Vamp

Winds of Change

Our world is undergoing a revolution right now. At the time of writing, it is hitting people in their pockets that something is changing. People are confused and don't know what to do or how to interpret their world. The changes that are happening now are part of a wider revolution in society that has been growing stronger over many years. We are gradually making a shift from one way of thinking to another. This shift will affect every area of our lives, including the way in which we view the concept of genius.

As we have already discussed, the current view of what we think of as a genius in our society is someone who is very good at using their mind like a computer, such as a very good mathematician for example. We also think of people who are exceptionally creative in skills such as art or music. There is a commonly held belief that only a few people are born with brilliance and the rest of us are just mediocre because we don't have the right brain or genes.

According to the current thinking, genius is caused by genetics and results in a high intelligence that can be rated by assessing an Intelligence Quotient or IQ. For those of us who have not been born with the right genes to create a high IQ, we believe we are not capable of anything exceptional.

Kraftwerk

This view of genius has it roots in the predominant scientific view of the last few centuries that sees the world as a machine. This is sometimes called the Newtonian or Cartesian viewpoint, after two of its main proponents. According to this view, the world around us is made up of smaller and smaller parts of a machine a bit like a clockwork mechanism. By understanding all the parts of the machine, you can understand all there is to know about the whole. This also applies to ourselves and our organs, including our brains.

Our brains are currently seen as machines that have parts called neurons which are somehow responsible for creating our thoughts. The mechanism of how this happens is not entirely clear yet, but it is believed that it will eventually emerge as science progresses. According to the current scientific paradigm, our thoughts are merely side-effects of the workings of our neurons. They are almost seen as an uninteresting fluke of an organ that just happened to grow very big through a process of evolution. Other species who do not have such large brains are often believed to have no capacity for thought.

And just as, according to current science, humans have superiority over animals, so are other humans capable of greater feats than the norm. These people are the ones we call geniuses. They, through the random, fluke processes of natural selection, have been endowed with the lucky hand. No matter that sometimes the emotional life of the person may not be fulfilling or they may even have disabilities; their brain is prized as an efficient machine. This is currently a common view of genius – as a bigger and better brain computer in a world of human biological machines.

This scientific viewpoint of the past few centuries that has dominated the Western World, has also managed to infiltrate countries whose indigenous philosophies originally did not see men as machines. Through processes such as colonization, the indigenous philosophies are all but forgotten and around the world, universities teach mechanistic science and therefore a biology which sees humans as machines.

So really we could call the mechanistic view the predominant global paradigm. The trouble with this paradigm is that it does not convey the full picture of reality. If all there was to an organism was its constituent parts, then we should be able to take bits of a cell, place them together and they would work together automatically. When you have a car that needs a replacement part, you can replace that part with another that may even be from a different manufacturer and it will still work. However, biological systems are proving more complex. It seems that even if you know all the parts of the 'machine', you still do not know all about the system.

Hunting High and Low

As I have said previously, the brain has also been seen as a machine, a bit like a sophisticated computer. The hope is that one day we will replicate this computer and create robots endowed with 'artificial intelligence'.[1] Now all we have to do is examine all the parts of the machine and then we shall have the secrets to intelligence and consciousness itself. IBM are currently spending a lot of money trying to 'reverse engineer' the brain.[2] This is the current mindset of neuroscience: analyze all the parts of the machine in order to understand the whole.

Except when it comes to the brain this isn't proving to be easy. It seems that the brain doesn't quite work like this although even people who are trained in biology still cling to the idea. When I was working as a GP, I attended a lecture at a hospital local to my practice, which was given by a neuroscientist. It was well attended by the local consultants and general practitioners, despite being out of working hours.

The attendees were asked to write down their questions for the speaker before he began. The top question for the speaker was, 'which part of the brain is responsible for consciousness?' revealing our curiosity to pin down a physical part to function, as we do in machines. The neuroscientist had to explain to the baffled crowd that actually, it has proved impossible to assign a part of the brain to the function of consciousness.

As Professor Susan Greenfield said in a lecture at the Royal Institution in London in 2008, "The brain has no lump on it that is consciousness".[3] The desire of the current scientific paradigm is to assign a function to each part including the quality of consciousness, but the brain is evading this type of pigeon-holing.

In fact, current neuroscience still does not know what causes consciousness although there are many theories available.

The philosopher David Chalmers coined the term, 'the hard problem of consciousness' to describe the thorny issue of explaining how our inner experiences can arise from a group of neurons firing electrochemical signals at each other.[4] How do these processes, complicated though they may be, explain our memories, thoughts and dreams? Nobody seems to have a satisfactory explanation within our current understanding of neuroscience.

We are beginning to understand something of the so-called 'easy' problem of consciousness which is highly complex in itself. The easy problem lies in explaining how all the parts of the brain are related to certain functions. For this purpose, neuroscientists have been using modern brain scanning techniques to examine what happens to the brain in certain situations. However, these scans don't explain *why* the brain lights up when we think about certain things, they just show us that it does. There is also no adequate explanation in mechanistic science as to how these neuronal firings are related to our thoughts and memories.

Senses Working Overtime

It seems that we can describe what happens when we think - the parts of the brain that light up on a scan, but not how we think - why these activities relate to our inner experiences. No matter how much we examine the brain, the answer still eludes us.

In fact the whole issue of consciousness is becoming a real thorn in the side of modern neuroscience, although it is easy to miss amongst all the headlines describing the next break-through about the brain. We have to keep in mind that all these studies are examining the effects that consciousness has on the brain and not actually examining consciousness itself. There are many research grants awarded to people who are carefully categorizing which parts of the brain can be associated with certain processes, which is a difficult task as the brain is hugely complex.

This type of research has resulted in, for example, finding out what happens to the brain when memories are created. We can track changes in neuronal structure that occur when short or long term memories are formed.[5] However, why something becomes a long term or short term memory is not clear. Nor is it clear how those neuronal changes become the types of inner experiences that we all have and call our memories.[6] There are also a wealth of animal studies that have shown that even when most of the brain has been destroyed, the memory of a learned task still exists, leading to holographic brain theories.[7]

It is this sort of conundrum that is frustrating the current scientific paradigm when it comes to the subject of consciousness. The trouble is that when we operate under a mechanistic paradigm, all we can do is track mechanistic changes such as chemical changes in neurons that occur when memories are created. We do not know what stimulates all these mechanisms into action.

It seems there is something more to our thoughts and consciousness than being a neuronal machine. However, by using the methods employed by current neuroscience and its world-view we are not going to find out what that is, because the methods themselves are mechanistic. They are designed to examine the *effects* of consciousness and not the cause. When it comes to the scientific basis of consciousness, if we keep searching within the current paradigm, we will be forever running in circles looking for the ghost in the machine.

It seems that a new scientific paradigm is called for - one that goes beyond the mechanistic world view in order to examine the phenomena of consciousness. Luckily that science is just arriving.

A New Hope

Whilst the biological sciences have been busy continuing the scientific paradigm of machinery, in the world of physics it has been a different story. Of all the sciences, physics is seen as the 'hardest' science as it deals most directly with actual descriptions of the universe itself: what the universe is made of and its laws of behavior. The way in which physics describes the universe tends to inform all the other sciences as well as society as a whole. For example the concept of The Big Bang has spread to most of the population, beyond the world of physics.

So if we underwent a change in physics, this should inform a whole range of sciences including biology. This is exactly what is happening; physics is indeed changing and the effects are trickling into the rest of science. Surprisingly, though it is the general public who are embracing the changes the most, as we have witnessed with the success of films such as *What the Bleep Do We Know?*[8] The biological sciences are largely living with the old mechanistic paradigm. As the changes in physics gain ground, this will have a huge effect on the way in which we see biological processes.

The Big Shift

We have just discussed how the world of physics is starting to undergo a revolution and causing deep changes to the world of science and to wider aspects of society. Amazingly the revolution was brought about purely by accident. The world of physics was not looking for these changes, but the implications of its findings have been rattling our philosophical cages ever since.

As stated earlier, physics is the science most associated with looking at the actual nature of reality and in the late 19th century and early 20th century there were huge leaps forward in our understanding of atoms, which at the time were thought to be the smallest building blocks of reality.

ATOM

When the scientists of that era looked more deeply into the nature of the atom they were in for a few surprises. First of all, atoms were not sold blocks at all, they consisted of a lot of empty space. So although the world around us appears to be solid, at this deeper level, reality turns out to be quite nebulous.

Another big shock came with the era of quantum physics. A number of experiments, which to this day have not been refuted, showed that particles of light called photons behave differently according to how we look at them; they change their behavior according to our behavior.

This was a major shake up to our view of the world. Before quantum physics, it had been assumed that we somehow stood apart from our universe and that it was possible to objectively witness our reality. We used to think that it was possible to perform an experiment without influencing it with our own behavior. The universe was 'out there' and we were simply observers.

It was a big shift to realize that the way we do an experiment affects the way a particle behaves. Until the early twentieth century, we thought of reality as being solid and indifferent to us. What quantum physics revealed is that sometimes particles behave as waves and sometimes as point particles depending on how we look at them. Particles have both wave and particle characteristics and we have a hand in determining which traits are displayed.

These amazing revelations about reality have been extraordinarily useful to us and the quantum age has been one of rapidly expanding technology. Quantum theory has given rise to innovations such as microprocessors and lasers, therefore it is involved in most of the technology we use today from mobile phones to PCs to CDs.

Most of us take advantage of these innovations without really thinking about the deeper aspects of quantum physics, but by using the technologies that quantum physics has enabled, we should also face the more philosophical issues which arise from it regarding the nature of reality.

WAVE - PARTICLE DUALITY

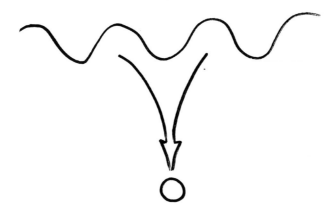

We can no longer think of the world as solid, and being made up of solid particles. We cannot even think of ourselves as apart from our universe anymore as our experiments have shown us that our own observations actually change the outcome.

It has been 'business as usual' for a lot of scientists, despite the revolutionary nature of this shift. In fact the whole arena of biological sciences including medicine, does not even take quantum theory into account and, on the whole, still operates under the older, mechanical model of the universe.

Yet the quantum revolution has occurred and there is no going back. Even though a lot of scientists wish to ignore the weirder aspects of quantum physics, those who are more open-minded actually embrace these strange ideas. Whereas many scientists find the concept of the involvement of an experimenter in wave-particle behavior as troublesome and wish this so-called 'measurement problem' would go away, there are some who have started thinking that maybe the fact that our observations of an experiment changes its outcome says something interesting about our own perceptions and their relation to the very fabric of reality.[9]

Brilliant Mind

Physicists such as Amit Goswami and Peter Russell have concluded that because it takes the act of observation to actually collapse a particle from a wave to a particle, there must be something intriguing about the act of observation.[10, 11] Amit Goswami states that in order for an experiment to be observed, a conscious entity must be present i.e somebody witnessing it. There is something about consciousness that interacts with the experiment and a deeper aspect of reality.

In fact Goswami goes further: he says that 'consciousness is the ground of all being'.[12] This is his logical conclusion from reviewing the many experiments involving the behavior of particles and how they respond to our own observations. There must be something about reality that has consciousness interwoven into it; in fact it must be fundamental.

This is a very radical statement. It means that matter emerges from consciousness not the other way round. That means that everything from your dog, a tree and your mobile phone arises from the same fundamental underlying consciousness as you do. You arise from thought - one could say from the Mind of God. It is no wonder that the ultimate computer in Douglas Adams' *The Hitchhiker's Guide to the Galaxy* is called Deep Thought![13]

CONSCIOUSNESS AS FUNDAMENTAL

The idea that matter arises from consciousness is the exact opposite of the materialistic paradigm which says that consciousness emerges as a by-product of physical processes in the brain. The field of neuroscience is plowing a lot of money into proving that consciousness arises out of the activities of neurons.

But as Goswami points out, this creates a paradox.[14] The brain itself is also made up of atoms and subatomic particles that behave in a quantum fashion. If they are collapsed from a wave into a particle by the very act of observation, by our consciousness, then what came first, the brain or consciousness? And herein lies the paradox.

However, as Goswami and others go on to conclude, there must be something fundamental about consciousness to the nature of reality and that, in fact, it is the 'ground of all being'.[15] It lies deep to everything that is a part of our universe, every atom, molecule and living cell and object all arise from an underlying sea of consciousness. It is in this way that the paradox is resolved as the brain's very structure also consists of consciousness, as it is made up of atoms.

Many spiritual traditions have discussed a similar concept over the centuries, saying that everything we see around us comes from one unified source - the Mind of God, Brahma or the unified whole. Mystics gain this sense of unity through their direct experiences of nature - some have written about these encounters. The trouble is, if someone has not had a unity experience themselves, they are unable to even comprehend that they exist. Hence the misunderstanding of mystics throughout history from Socrates to Joan of Arc.

It is ironic, therefore, that it is via modern science that we are coming to the same conclusions as mystics throughout the ages; that all the varieties of form that we see around us emanate from one source and that source is consciousness.

Even though modern science is being interpreted this way by some people, this is certainly not the majority view amongst scientists. In fact, there are many who are vehemently opposed to it. Yet there is a growing movement in science that is starting to accept the deeper implications of quantum physics and other areas of modern science and, with the popularity of books and films in this genre, the public is embracing this view too.

It seems that although the decline of religion has been sustained over many years, this does not mean that people are now devoid of spirituality. In fact the opposite is true. Instead of seeing God as a person who sits in the sky overlooking all of our deeds, people are embracing the idea that there is a divine intelligence and cosmic order - more in keeping with the ancient Eastern beliefs. (In fact these views have been found all over the world, but are more associated with places like India and religions like Buddhism.)

This view of the universe completely changes our notion of what intelligence and genius is. From genius being the result of a fluke in a mechanistic brain, we can now see that according to modern science, the universe is alive with intelligence and that consciousness is not just found in one place, the human brain, but deep to the heart of everything. What happens when we realize that we are all bathing in a sea of intelligence? We need a new user's guide to the brain.

Chapter 5

The New User's guide to the Brain

"There's been a brainwave at the radio station

Old idea from the Woodstock generation

Calling all the kids from across the nation

In some it brings out love, in others termination"ʃ

Big Audio Dynamite

<u>Half a person</u>

When I was younger I used to try and defrost my brain. You see I had this strange idea that half of my brain was somehow frozen since birth and if only I could melt it then I would suddenly become superbly intelligent, beat everyone at exams, get the best job and live happily ever after.

I had bought into the paradigm, even as a young child, that the machinery of your brain is responsible for your intelligence and I thought that the more brain I had, the more intelligent I would become and the more successful I would be. I must point out that I was brought up by parents who were both doctors, including a mother who had topped exams in both her school and medical school. Her focus for us as children was very much on educational success and passing exams as the route to a better life. Hence my early beliefs were that examinations are the key to success. I have since learnt that education is only a part of life, of course.

I am sure that many people today have the same belief - that if you happen to be endowed with a large brain or one with better machinery, you are more likely to be a genius. Just look at how we peer at Einstein's brain which has been preserved since his death hoping to somehow gain some insight into the secrets of genius. (More on Einstein's brain later.)

This view that brain equals intelligence and therefore more brain equals more intelligence fits very nicely with the mechanistic reductionist paradigm and also with the notion that all there is to know about the world is measurable and physical. The latter is an assumption that is so widespread in both society and science that it is hardly even mentioned.

So when we hear about certain people who do not have much brain mass, but are capable of living normal lives and even achieving great feats of intelligence, this challenges this world view and our own personal assumptions. How can people with little brain mass be capable of achieving anything at all? Yet there are recorded cases where this has happened.

For example, there is a lady from the UK called Sharon Parker, who was featured in a British TV documentary in 2003, who has very little brain mass.[1] She suffered with hydrocephalus as a child which is a build up of fluid in the brain. This fluid squashes the actual brain matter inside the skull. By the time the condition was detected and treated, there was so much fluid in her brain that Sharon had only 10-15% of the normal brain mass. Her parents were told to expect that she would suffer from disabilities.

In Sharon's case however, it seems that the fluid built up so slowly that the brain was able to compensate and, despite the dire medical predictions, Sharon has led a relatively normal life and takes an active role in her husband's business taking care of his accounts.

And there are other recorded cases that are quite baffling if viewed from the current scientific perspective. There are a few examples of people having very little brain mass due to diseases such as hydrocephalus and gaining degrees in mathematics.[2]

Sometimes people can perform feats that should be impossible under the current paradigm's understanding of the brain. There are many recorded cases of a phenomena known as Blindsight.[3] This can occur in people whose optic tracts of the brain are damaged to such a degree that they should not be able to perceive visual stimuli. However, under controlled tests they are still able to demonstrate that they can respond to visual stimuli. Somehow, some people who are 'blind' are able to 'see' by a mechanism other than the ones known about in the current scientific paradigm.

What exactly is going on? These phenomena seem mysterious if you are still stuck in the old paradigm of seeing the brain as a physical machine which produces consciousness. But we are moving away from that paradigm into the realm of the New Science in which atoms are not so solid and consciousness is fundamental to matter. In the new world, we can be open to novel explanations of how the brain works and therefore the true mechanism behind genius.

It's like atomic theory

Let's recap on some of the ideas of the last chapter. We discussed earlier how, at the most fundamental level, the universe is not made up of solid particles as scientists originally thought.

These particles make up the 'stuff' we see around us, which means this is also not solid. This means the brain, which is also made up of atoms that display quantum behavior at a deep level.

If the brain is made up of atoms that display quantum behavior, this would also include wave-particle duality. This means that even the atoms of the brain collapse from a wave into a particle when observed. Some argue that this requires a conscious entity to observe the particle to actually collapse it into a wave.

But if the brain is truly the agent that creates consciousness, which is the argument of neuroscience as it stands right now, then how is it that the atoms of the brain exist in the first place? These atoms need consciousness to collapse them too. To make things worse, according to some scientific viewpoints, human beings are the only animals considered capable of true consciousness, because they have a highly developed forebrain.

How did all these atoms exist before human brains developed the capability of collapsing all these waves into particles and then bringing things into existence? Surely there is another answer. As I said earlier some physicists are now coming to the conclusion that this paradox only makes sense if we understand that consciousness is fundamental to the universe and that all matter emanates from a deep underlying 'ground of all being'.

The modern scientific view seems to be converging with the old view of ancient mysticism - that we live in a world created from a universal intelligence and this pervades everything we see around us. What is interesting is that physicists have come to this conclusion from a purely logical perspective.

So we are bathed in a sea of intelligence: it interacts with us, it unites us, it creates us, it is who we are at a fundamental level. We live in an illusion that we are separate beings distinct from one another and from the source. Our path as human beings is to realize this and this process of awareness is sometimes called enlightenment and self-realization. The transcendence of the limitations of the self to realize that every being is at one with an underlying order is not an intellectual process; it is fully experiential, but often indescribable through our language constructs. Often, it is only when someone has had such an experience that they understand it. Otherwise this sort of concept is unconceivable to a person whose awareness remains in the realms of the separate, material reality.

Invisible Touch

What is happening right now to humanity is that an increasing amount of people are experiencing enlightenment and self-realization. This is leading to a changing view of the universe, and different way of living. Our sensory perceptions are changing, including our sense of time and space. People are becoming more in touch with realms that are hidden from normal perception. Maybe they are afraid of talking to others about it, because the only way of viewing these experiences in terms of current psychiatry is as a mental illness that needs treating. But the fact remains that these experiences are increasing as if we are being 'rewired' for different modalities of sensory perceptions.

Now that our perceptions are expanding, the old mechanistic ways of viewing intelligence are no longer adequate. We are moving into a new view of the universe with a different way of viewing consciousness, intelligence and genius. Whereas before, intelligence was seen as the product of machinations of the brain, there are scientists who are realizing that intelligence does not end with the neuron; the whole universe is intelligent. In this new view of intelligence, thoughts and consciousness come through the brain not from it. The brain is a channel for consciousness not the source of it.

This is why no matter how much we take the brain apart, we cannot find out why neurons are able to cause the inner experiences that we all have. Some philosophers such as Daniel Dennett believe that the inner conscious experience simply emerges when enough neurons are gathered together to form a human brain.[4] Yet there is a huge logical inconsistency in this argument. Although some systems in Nature do indeed show emergent properties, when it comes to our complex inner experiences it seems trite to simply explain them away by saying that if you get enough neurons together something magical happens.

Nobody has satisfactorily been able to explain how this emergence happens; if one neuron does not display any magical properties then why and how do a whole bunch of them? It makes much more sense to describe consciousness as already existing in the universe: you do not have to magically create it from a bunch of neurons. Consciousness *is* everything and is everywhere with no special circumstances involved. This is simply a logical conclusion of quantum physics by some quantum physicists.

So we now have a new explanation for how we are able to have thoughts at all. Our brains do not create consciousness, but instead are mediums to convert the consciousness that exists in the universe itself into our own personal thoughts. The thoughts exist in the sea of consciousness that surrounds us - we are merely conduits. This is the radical new view of consciousness and our brains that is arising out of the scientific revolution.

The Quantum Sea

We have looked at how currently, scientists cannot explain the mystery of consciousness. Some do not believe that it exists at all and wish to ignore the phenomenon. We have also examined how there are some explanations of consciousness arising that are more radical than the reductionist view.

Some have tried to incorporate the science of quantum theory into their theories of consciousness such as Cambridge mathematical physicist, Sir Roger Penrose and Arizona-based anesthetist Stuart Hameroff.[5] These theories utilize the uncertainty effects in quantum theory to postulate that these strange effects give rise to our inner experience of consciousness. These types of theories are a sort of half way point between the reductionist and the new emerging world views. However, they seem to be inadequate for explaining Blindsight and other such phenomena.

There is a new emerging vision of the way the brain works that better explains these phenomena. To understand it, we have to revisit the principles of quantum theory. As stated earlier, one of the major surprises in quantum theory was the concept of wave-particle duality - that particles exists as a wave of possibility until they are observed, which is when they become point-like particles. Until they go through this collapsing process, they are in many states at once - physicists call this *superposition*. This implies that a particle is in a state of possibility and has a chance of existing in many different states and even places at once. According to statistical probabilities, there are infinite possibilities of where a particle is located.

From this idea, the early quantum physicists realized that a particle could be popping in and out of existence at different locations. So a particle that exists in the Eiffel Tower also can exist in the Empire States Building at the same time - physics allows for this.

It also means that there is a field of 'virtual' particles around us to reflect this statistical potential. Experiments have been done to show that this is indeed the case. There are particles that exist for a fleeting moment before disappearing. This is occurring all around us.

This field of virtual particles is called the Quantum Vacuum or the Zero Point Field. It consists of particles of light called photons which split into two particles of matter and antimatter for a fleeting moment before becoming whole again. Antimatter and matter are like mirror images of each other. This continuous cycling is happening around us all the time - we are bathed in it and it is active even in a vacuum and approaching zero point temperature, hence the names.

What is also interesting is that a photon can travel at the speed of light, it does not have mass or charge, whereas the particles of matter and antimatter do have charge and mass. So charge and mass have been created out of 'nowhere', they fleetingly exist before disappearing again. This charge that is created all around us, even exists in a vacuum and the effects of this charge can even be measured - they are called Casimir forces.[6]

We are surrounded by a sea of light that is constantly changing and becoming particles of matter and antimatter before transforming back into light again. It is a bit like the ouroboros snake chasing its tale: a continual death and rebirth process.

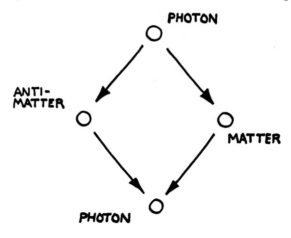

The Holographic Principle

So what does this have to do with genius? Well another way in which we can look at a particle is as a wave of information. As a particle can be viewed as both a wave and a point particle, it follows that we could also view the Quantum Vacuum as a sea of waves of information. We know that when waves cross and intersect, they store information in their intersection points - we use this principle in our everyday laser technology. This means that according to physics, we are surrounded by a sea of information, stored in the intersection points of the waves of light that surround us.

This sea of information is also *holographic* which means that one part contains the information of the whole. An everyday example of a hologram is displayed on most credit cards: the shiny picture that is there to prevent fraud.

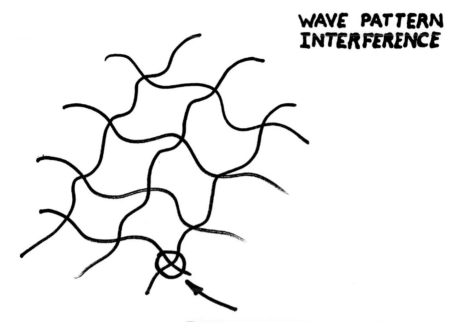

WAVE PATTERN INTERFERENCE

INFORMATION IS HELD AT INTERSECTION POINTS

Another example of a hologram is the image of Princess Leia seen in the *Star Wars* movie, which is a three dimensional image that looks like the object but is not the object itself.[7] Holographic communication technologies are actually in use today via companies such as Musion.[8]

A simple hologram can be created by bouncing laser light off the object then crossing it with another laser then putting those crossed beams through a photographic film. The film then gains a swirly pattern which doesn't look very interesting. However, when you shine another laser through this film, an image appears of the original object that seems three dimensional in nature.

What's more - the film can be cut into little pieces and each piece will still produce the image of the whole object. The concept of each piece containing the information of the whole is known as the *holographic principle*.

CREATION OF A HOLOGRAPHIC IMAGE

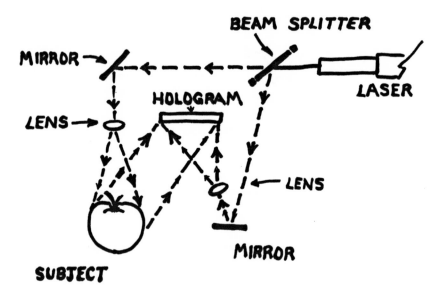

The universe as a whole is being seen by physicists as a hologram, both *New Scientist* and *Scientific American* have reported this concept as cover stories.[9, 10] If the universe displays the qualities of a hologram we can view it as a sea of information where each part of the universe contains the information of the whole. That means that everything that has ever happened, is happening or will happen to any being or anything is stored in the holographic field.

So now we have a storage site for the universal intelligence that pervades our world - it is the Quantum Vacuum. Furthermore, we interact with this universal intelligence. We have said earlier that the brain is not the source of consciousness - consciousness is the ground of all being. We have also discussed how modern neuroscientists cannot explain how we have inner experiences. We can see the changes in the brain that memories make, but we do not know what stimulates the change. Try as we might to search the physical realm, we cannot find the ultimate stimulus - consciousness. Could it be that consciousness does not exist in the brain at all?

In the new scientific paradigm, the brain becomes a conduit for information from the Quantum Vacuum, which is information stored in light waves. The intersection points of waves are the most efficient storage mechanism we know of. In fact researchers such as Karl Pribram have already postulated that the brain works as a hologram and uses the intersections of light to store memories etc.[11] This was after some animal experiments that found that you can destroy many parts of the brain without destroying memory, implying that memories are not stored in a particular place in the brain.

Let's Get Together

We now have all the pieces in place for understanding the New Science of genius.

- We have looked at atomic theory and discovered that seemingly solid objects are actually mostly empty space.

- We have examined quantum theory and discovered wave-particle duality that says that a particle exists as a wave of possibility until you observe it.

- We have seen how some physicists think that consciousness is necessary to collapse the atom from a wave.

- We have seen that some physicists conclude from this that consciousness is fundamental to reality and pervades all things, every atom and molecule - that it is the ground of all being.

- We have understood that there exists around us something called the Quantum Vacuum or Zero Point Field which can be viewed in two ways. In one way, it consists of many photon particles of light that continuously split into particles of anti-matter and their opposite matter. In another sense, it can be viewed as a field of light waves of information.

- This information could be seen as a storage mechanism for the consciousness that pervades our universe.

- This field is also holographic so each part contains the information of the whole.

The new scientific view of consciousness and the brain follows on from these findings and says that the brain transmits information from the Quantum Vacuum. It is consciousness itself that is lighting up our brains when we have a thought.

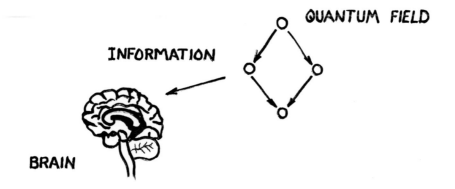

Consciousness is not generated as an epiphenomenon or side effect of our brains, it is everywhere and our brains are simply the conduit for it. This is the new view of how the brain is activated, but this picture does not fit the old mechanistic scientific paradigm which cannot recognize anything that is non-physical. Universal consciousness is the 'ghost in the machine' that neuroscientists are tearing the place apart looking for. Once science has undergone this current paradigm shift, we can have greater understanding of the brain and also of the true nature of genius.

This 'consciousness-is-everywhere' paradigm can also explain why bacteria and single-celled organisms without brains, show signs of intelligent behavior. In fact every one of our cells in our body is highly organized and shows signs of intelligent behavior, but none of them have brains. This is because brains are not necessary for consciousness or intelligence. As with any conduit, if the brain becomes damaged, by a stroke for example, this will affect the translation of consciousness through it. But there is evidence that people in a coma can remember what has been said to them and people have even had conscious experiences when their brains are showing no clinical activity at all.[12,13] The brain is not the source of consciousness; consciousness exists beyond the brain.

The New Genius

Now we have recapped what we have learnt, we can apply this knowledge to understanding genius. We have come far from the view of 'more brain means more intelligence', despite my youthful fantasies of melting half my brain to unleash its supposed potential. Thoughts are held in an information field of light and are able to come through the brain for them to be articulated or translated into movements.

Our brains, indeed our entire bodies, can be seen as a translation tool for the universal consciousness for us to process and make use of in our lives. Our thoughts totally connect us with our universe. Through the quantum field and its holographic sea of information we have access to infinite knowledge.

And therein lies our genius. Each of us relates to the Quantum Vacuum in a unique way. We can bring information through from the Quantum Vacuum and into our lives that is unique to us. The more we practice this, the more we get into our zone or Genius Groove. This is when we are no longer as aware of space and time and able to get deeper into the field. This is when those moments of inspiration can come through easily. By tapping deeper into the field we are actually tuning into our own consciousness at a deeper level.

This is how we get those moments of inspiration that come when we are not looking or trying, when we are able to let go. Sleep is a common way to let go and we often hear people say that they will sleep on an issue. One of the world's most successful fashion designers, Karl Lagerfeld claims to wake up with new designs in his mind - this is just one example.[14]

By moving our consciousness further out of the space-time realm, through meditation for example or going for a walk, we can tune into the Quantum Vacuum at a deeper level and gain information that reflects more of who we are. We become less like the antimatter and matter, worldly side of the field and more like the uncharged, photon side of the field: out of space and time.

This is the New Science of genius and it is gaining ground over the old mechanistic, IQ-based view of genius through the work of scientists and authors such as Ervin Lazslo, Michael Talbot and Lynne McTaggart[15, 16, 17]. Even though we all have our own unique genius and unique way of relating to the universe and the Quantum Vacuum, for various reasons we have been taught not to practice tapping into the field and our own creative source. As previously discussed, we live in a system that has been created in order to keep us disempowered and following an authority outside ourselves. As we start to tap into the field and step into our Genius Groove, we can break free from the conditioning that has been holding us back.

Once you realize that genius isn't something 'out there' given to a few gifted people and that everyone is a genius in their own right, you can begin to see that you were a genius all along.

As we start to understand the New Science of the brain and consciousness, we can begin to understand phenomena that previously had no scientific explanation. In fact the previous explanation of the brain as a machine does not stand up to closer scrutiny. It does not make sense of some of the talents of various people, for example the psychic mediums, found in every culture and civilizations throughout humanity, people with very little brain mass who have degrees in mathematics or Blindsight and other special abilities.

Some people can access knowledge that was previously thought as impossible - that transcends space and time. These people are sometimes call seers, mystics or psychics. The old scientific paradigm has no explanation why the behaviors displayed by psychics appear in every single culture and every single era of humanity. If there was no basis for psychic mediumship, other than a few people acting as charlatans, this behavior would not be so pervasive. If one accepts that these people are simply adept at taking information from the field and translating it through their brains, their talents start to make sense scientifically.

In fact many studies show significant results when it comes to testing psychic abilities.[18] Some have tried to dismiss these results because they simply do not fit in with their world view, but now we have a scientific explanation for these phenomena.

Oscillate Wildly

Indeed, some of these experiences fit with another theory in mainstream physics - String theory. This aspect of physics was born out of the need to make things neat and beautiful in physics. We currently know of lots of different particles and forces that don't tie into one unifying idea. String theory describes all of these forces and particles as tiny vibrations. We can never hope to see these vibrations - they are too small. This theory has emerges from mathematical equations, not experiments.[19]

Furthermore, these strings need many more dimensions of space and time than we are used to. Our world consists of three spatial dimensions; side to side, back and forth and up and down; and also a dimension of time. Hence we speak of meeting someone on a King street, on the third floor (spatial dimensions) at three o'clock (time). These are all the dimensions we need to locate someone in our world.

We are normally incapable of visualizing higher dimensions, but according to some quantum physicists, consciousness is fundamental to the universe and unites everything in the cosmos. So could it be that mystics throughout the ages have been accessing higher dimensions of consciousness? That might explain why predictions made by prophets such as Edgar Cayce have come true.[20] Was he so talented at accessing the holographic field of information that he accessed another dimension?

String theory also gives us an explanation as to why we each relate to the field in our own unique way. We too are made up of vibrations. Each of our atoms have their own vibrationary signature as do the combination of all our atoms. Vibrations have resonance, a bit like music. Resonances that are similar attract each other, as explained by the science of Sympathetic Vibrationary Physics.[21] Our unique personal vibration attracts information from the field which best matches our unique signature. This gives us each our unique genius.

If everything in the universe is made up of vibrations in many dimensions which are fundamentally conscious then we are living in a multidimensional, resonant world - something Michio Kaku calls the 'music of hyperspace'.[22] Our own consciousness resonates and attracts like vibrations - hence people end up repeating the same sorts of relationships and jobs despite wanting to change. There is a way to change your consciousness vibrations and that is through emotional resolution and spiritual growth. We will discuss the Law of Attraction and emotional resolution in more details later on in this book.

Pi in the Sky

Phenomena such as Blindsight can be better explained when we understand that the brain and even the body is a conduit for consciousness that exists everywhere. The optic tract, therefore does not have to be physically exposed to a visual stimulus in order for a person to be able to interpret it.[23] The stimulus can be conducted by another part of the person from the universal consciousness.

Daniel Tammet came to my attention due to a British television documentary, but he has since written and published his autobiography, *Born on a Blue Day*.[24, 25] Daniel has been labelled as an autistic savant because of his amazing mathematical abilities; he even learns languages very quickly too. So far, he sounds like the same sort of person we were talking about earlier, the ones who happen to have been blessed with a brain that seemingly works like a human machine.

However, Daniel has several interesting qualities. First, he was not always gifted so strongly in this way. He had a epileptic seizure as a child which triggered an unusual way of viewing the world. After the seizure, Daniel began to see the world in numbers. Numbers appear to him as shapes which gradually coalesce. He recognizes each number by its character. Neuroscientists have tested him to find that, unlike other savant subjects they have studied, Daniel is not calculating like a human computer - he really is seeing numbers as shapes. They apparently come towards him and take form. This is how he is able to give the answers to complex mathematical problems so quickly.

One of the interesting parts of the program was when one of the researchers wrote down the number pi to a certain number of decimal places, but changed some of the numbers to test if Daniel would notice. Daniel did indeed notice and felt disturbed and complained that something terrible had been done to something so beautiful, which was the number pi as he sees it. The neuroscientists in the program were baffled by this and called for a new science to understand how Daniel is able to behave the way he does.

Although this baffles current neuroscience, it fits our Quantum Vacuum picture of genius because information is drawn from the field and then we able to articulate it. Daniel is probably relating to the field in an unusual way, triggered by his childhood epilepsy, but in doing so, he demonstrates the true mechanism of the brain as a conduit.

Daniel has a particular type of genius that allows him to see numbers and learn languages very quickly, but we all have our own type of genius. For all of us, information is continually coalescing from the field - these are our normal thoughts and memories. As we go deeper into the field, we uncover more and more of our unique Genius Groove. Everyone has their own journey. The rest of this book may help you find your inner creativity and get into your Genius Groove, but before we leave the subject of the brain, let's examine the curious case of Einstein's brain.

Einstein a Go-Go

One of the strangest episodes in scientific history is the story of Albert Einstein's brain. As was the custom at the time of his death in the 1950s, his brain was dissected at an autopsy so only photographs of the intact brain exist. For many years the sections of the brain were kept in the house of a pathologist involved with the autopsy until it was returned to one of Einstein's relatives.[26]

Of course, Einstein's brain has sparked scientific curiosity as we are fascinated to see if it reveals anything about the man's genius. A study in 1999 in Canada revealed that the parietal lobes of Einstein's brain were larger than average.[27] Interestingly, the types of cells in these areas are different from control groups; Einstein had a higher glial cell to neuron ratio.

Most people know about neurons in the brain, but might not know about their less studied partners - glial cells. Until recently these cells, with a higher fatty content, were thought to have a purely supportive role, but modern imaging techniques show a very different picture revealing complex chemical communications between glial cells which are still poorly understood.[28] Interestingly, glial cells can continue dividing longer than neurons. We now know that the brain is a plastic structure that can adapt over time and to special circumstances eg. London cab drivers display changes in the hippocampus region of the brain.[29]

Could the glial cells be responding to the way in which we take consciousness from the quantum field? Could Einstein's brain have shown adaptive changes because it was a particularly developed conductor of information? Pathological liars have been found to have an increase in white matter in certain regions of the brain, which again is an increase in fatty material.[30] Does this reflect a well-developed imagination?

Children who have suffered severe neglect from birth with nobody to teach them language, display a noticeably different brain structure on MRI scans.[31] Taking in information seems to actually mould the structure of our brains.

It is worth considering that the structural neurological changes we see in humans who are focused on a particular activity are due to better transmitter function. Until more work on the brain is done, nothing can be concluded. However, it may be that getting into your Genius Groove actually makes your brain a better conductor of information!

The fact that we are discussing this topic at all is a revolution in science as not that long ago, as late as the 1990s, we thought the neurological system was unable to adapt. This view fitted with the old paradigm idea that viewed the body as a machine.

The evidence from Einstein's brain and other studies is leading us to realize that something is able to change the morphology of the brain, which adapts to circumstances. Again we need to look deep to the atom and to incorporate Dr Bruce Lipton's work to see that there is something outside the cell that is making those changes.

The conclusions of both quantum physics and the new biology are that it is not the structure of our brains that affects how we think, but where we place our attention and how we perceive the world that alters the brain.

So as we get deeper into our Genius Groove, by developing a relationship with our particular vibration in the Quantum Vacuum, new pathways are developed in our physical structures that make getting into the groove easier the more we practice.

Now that have examined the New Science behind creativity and The Genius Groove, let's look at the steps we need to take in our own lives to get there. I can think of no better place to start than telling my own story in order to illustrate these principles and how I came to learn them.

Part Three - Live to Tell

Chapter 6

Borderline

"Black my story,

it's not history" [8]

Ziggy Marley

Now you have the science behind The Genius Groove, it is time to find out how to get into it. To do this, I am going to share with you the story of how I got into mine. My story is harrowing as well as inspirational. It is my version of events and the people who have been key players in my life might remember things very differently...

Back to my Roots

I was born in Yorkshire in the UK to parents who were both doctors by profession. My Dad was the first son of a doctor in India and had decided, when his father died, to get out and see the world. Otherwise he feared he would be stuck in Kolkata following in his father's footsteps.

He came to the UK in 1972 originally intending to move on eventually to the USA. My mother soon came to join him with my elder sister who was three years old. My mother had been a star pupil at my parents' medical school and was tipped for the top of the profession. She left India without formally resigning her job as she knew that her resignation would not be accepted, such were the high hopes for her career.

They worked their way up from humble beginnings; when they arrived in the UK they lived in just one room in Leeds in the North of England. When I came along, they eventually moved into their first house. My younger sister arrived and the houses became bigger. My father never quite lost the travel bug and we would visit India, Canada and the USA as well as seeing all the sights in the UK.

I suffered, as do a lot of middle children, from being in the middle: not being distinguished as the eldest or youngest. My elder sister dominated the household. I recall that she did a share of parenting, as my parents were so busy in their careers.

We had various child minders over the years as my parents tried to juggle their work with having three children. This was fairly unusual at the time as women were still expected to give up work and stay at home with the children.

I attended a private day school from two and a half years old and continued to have a good private education until my graduation at the age of eighteen. From very early on, I sensed an expectation that I was to become a doctor. My parents even bought me a doll's hospital to encourage me. People could see that my calm temperament was the most suited out of the three girls for a medical life and would stand around and saying things like, "put that one through medical school", pointing to me.

Our childhoods were very exam orientated. My mother never failed to miss an opportunity to tell us that she was a brilliant student and had won academic medals. We felt a lot of pressure, as many Bengali kids do, to perform. But we were also different from other Bengali kids - my elder sister had strong interests in a range of subjects including artistic subcultures. With discussions on topics ranging from Andy Warhol to

Vivienne Westwood and William Burroughs, our house was a cultural incubator. We all had a strong interest in fashion which was quite unusual at the time. Family trips to London often involved investigating the up-and-coming designers in Kensington's Hyper Hyper store for example.

As the pressure for exams increased in my teens, my cultural interests started dwindling. From being a ferocious reader, I stopped reading novels or taking guitar and dance lessons. I worked very hard at school, but somehow never reached the top; I just wasn't good enough. This added to my feelings of mediocrity - the middle child and the middle student. Despite this, I managed to get into medical school; my rounded interests probably tipped the balance in my admission.

Teenage Dirtbag

My older sister, 'B' was a very important figure in my life growing up. She dominated the whole family - what we chose to do, our tastes, our clothes. For some reason, I found myself afraid of her and I didn't know why. Sometimes she would come after me and threaten to hit me. I acted very meekly around her. She would often tease me in public and recruit other kids into the campaign, causing me a lot of pain and humiliation.

She left for university when I was fourteen and, after an initial shock, I really came out of my shell. I went from being a quiet, scowling 'goth' teenager, obsessed by indie music and dressing in layers of black, to being a bundle of positivity. I started playing a variety of sports and even won public speaking competitions. These were some of the happiest years of my life. I enjoyed school and home life, I was popular and had a purpose in life - to get into medical school.

When I left home at eighteen years old, I lived with my elder sister in the flat my parents had bought in London. My medical school was actually just across the road. The relationship with my sister began to deteriorate. She soon started to become violent against me. My grades started slipping. I turned to a boy I knew from the Bengali community - he became my first boyfriend.

In the middle of all this, I underwent a spiritual awakening which was quite unplanned. I had what is known as a *kundalini* experience which is a Sanskrit word for coiled energy that lays dormant in the sacrum bone until it is awakened, by meditation or other means. Suddenly, I received what is commonly known as enlightenment - I was plunged into union with the universal mind. This experience started me on a whole new path of spiritual discovery. So I underwent all this on top of attending medical school, which is hard enough on its own!

Moving to London felt as if all the pieces of my life had been thrown up into the air. The joyful, positive teenager that I had been started to slip away. I was losing my sense of self - between the boyfriend who was smoking cannabis and therefore making me passively stoned, to my elder sister's abusive behavior, my burgeoning spiritual journey and discovering that medical school was not a right fit for me - I was not in a good space and my university grades reflected this.

Papa Don't Preach

On top of everything, I became pregnant after an accident with a condom, despite asking my mother's advice regarding emergency contraception. I knew immediately that I couldn't keep the child. It was a strange feeling, I was connected to the soul inside me and the spirits around her, but I knew she was not to be born yet. I was highly traumatized by the whole experience - I felt guilty and cried every day for at least a year. My boyfriend and I decided not tell our families which made me feel even more alone.

One day, I was feeling bad and crying alone in my bedroom, I looked up and saw a beautiful shining angel about eight feet tall. I was filled with the overwhelming feeling of unconditional love and the sense that nothing you could do could ever be wrong - there is nothing to forgive and everything is perfect so there is no need to feel guilty at all. I had several such visits at the time and angelic connection is still a strong part of my life.

I realize now that the whole experience, though traumatic, was furthering my spiritual connection following on from my kundalini experience. My short pregnancy was an extremely multidimensional experience - something I would not have learnt in any other way at the time. Despite the angelic visitations, I continued to be depressed for many months afterwards.

Twisted Sister

After a few years of feeling this downwards escalation, I visited one of my old child minders and discovered why I was so timid as a child. She told me that my elder sister had repeatedly beaten me when I was very young. B used to ask this child minder to conceal this from my parents, which caused a lot of guilt in her. The violence I was experiencing at that moment was therefore a repeat of a pattern that had happened before. I was starting to understand why I behaved so submissively in life - a pattern I am still coming to terms with.

Gradually the violence escalated. A few times I found myself alone in the streets of London in the early hours of the morning. Sometimes, I would go to my boyfriend's flat or another friend's place. It was a very frightening time.

I asked my parents to put me into another flat and even found one to stay in, but they refused to pay the rent. I even went to the police to make a complaint to the local domestic violence unit, but my parents still would not do anything to halt the violence; I lived a life of constant intimidation.

It was when my sister made an attempt on my life that I really got a wake-up call. One day, a row erupted out of nowhere. She hit me hard against the kitchen window and then threatened to stab me. I ran away, but she was on the other side of my bedroom door with a knife. The weird thing was, part of me wanted her to kill me as I didn't think my life was worth anything. I felt so useless that I might as well die. I was only twenty years old.

In that moment, I became aware of my own thoughts and was shocked at what I truly thought of myself. It had taken this extreme situation to show me just how little self worth I had. For the first time in my life, I wanted to live - I thought I was worth it.

Run to You

Shortly after these incidents, I met the man who was to become my husband, 'J'. He was the top student of the medical school and was a little older than me. We had an instant connection and I was shown, to my surprise, many signals that I was to marry him. The universe eventually wore me down with the amount of signs I was sent. I had niggling doubts, but I pushed them aside, knowing that marrying him felt like the right thing to do. I had not finished medical school at this point, but the spiritual signals were so strong that I married him within a year of meeting him.

My parents opposed the marriage initially, but married life suited me; I settled down and my grades improved. I soon found out my husband was not as savvy as I thought he was, despite being a high flyer at college. He was a constant spender on gadgets and electronics. In those early days, our North London rent was high, yet he was bringing in very little salary and I was still a student. I could no longer afford the tube fare to go to college so even missed out on my final year photo. We got by with help from both our families which I am so grateful for.

We managed to buy a flat in North London and I passed my exams and became a junior doctor. My husband, however was struggling with life and did not enjoy clinical work even though he was extremely knowledgeable. I hated medicine too, but I did not shine at it like my husband. Every day at work was a struggle for me; I constantly felt I was stupid and in the wrong career.

The Greatest Love of All

I remember shortly into my marriage, feeling invisible and crying, "What am I good at? What am I good at?" I was sick of being mediocre. My husband's success only meant I became more invisible in people's eyes as they assumed he had tutored me through medical school (he had not). I hated the fact that I had gone through life working hard, but not achieving much when compared to others.

So I started a secret life: my spiritual quest. I started to read about the scientific basis for spirituality. Books would seem to fall off shelves by Deepak Chopra, Christiane Northrup - doctors like me who knew about the spiritual dimension of life. I also did a lot of healing of my emotions, reading numerous self-help books and coming to terms with the years of abuse and recognizing my own anger.

My secret life grew despite the fact that I was working long hours. I attended conferences and lectures in my time off - all without my husband. It soon became clear that we were growing apart. He was not interested in the spiritual dimensions of life and worse still, was no longer interested in me, as one of his college friends, whom he always held a torch for, had split up with her husband and now seemed to be a better option.

By the time I was training to be a GP, his career had hit a low point as he quit his prestigious medical rotation to do locum posts. There came a point when he no longer wanted to work in medicine and I became the sole breadwinner. By this time we had moved to St Albans in Hertfordshire from North London. I used the money from the sale of our flat to pay off all the debt we had accumulated. I also furthered my spiritual path, by training as a bio-energy healer. This served to drive my husband and I further apart as I become even more spiritually inclined.

In a strange series of events that seemed so meant to be, I got a job as a GP trainee in a nearby town of Berkhamsted. As I was about to do on-calls from home I thought it would be a good idea to move there. We also liked the place and I thought getting a new house together would be the answer to fixing my marriage. Once we moved in, J found a perfect job for him just round the corner. I finished my GP rotation in the new town then went straight into part-time work and started organizing local spiritual events.

I was realizing more and more that my husband and I were parting ways, but often I was too busy to really think about it. In 2002 I took a career break to try and gain a semblance of my true self. Once I had the time and space to think, I realized very strongly that my husband and I were no longer suited and should part ways. He was also spending more and more and time with his female friend from college and admitted to me that he could not choose between us.

Over the period of about six months, I talked things over with him, but he couldn't comprehend who I had become. I knew that my marriage had come to an end, but I just needed a push to get out.

Secret Lovers

During this time, someone came into my life to give me the jolt that I needed, let's call him 'L'. I had actually met him when he came into my surgery as a patient years before. Now I was no longer his doctor and, as my marriage faded, he started to take more prominence in my life. At first he joined my circle of friends, but there was always some spooky connection between us. I realized that we were some sort of soul mates, but I did not think for a second that we were suited as a relationship.

Sometimes, things can happen despite your best intentions. One morning when we were saying goodbye to each other, a wave of energy came over me. In a flash, I saw every lifetime that we had been together and every time I had seen him die in front of me. I had a strange feeling that I couldn't let him die in front of me again and before we knew it, we were having an affair. As my husband and I were already breaking up, it seemed fitting that I went to live in a house associated with the family of L and my husband live with his college friend.

I'm Going Down

However, when J was rejected by the college friend whom he had kept a flame for all these years, he suddenly exacted revenge and struck out at my career. You see, even though those days were harrowing for me, I had managed to hold down several roles to earn money - I worked as a holistic doctor at the Bristol Cancer Help Center, as a GP and also as a healer and had started giving talks on the subject of the science of healing.

J started to use my spirituality against me when speaking with my parents. He convinced them that my multidimensional life, my seeing angels and having visions etc, was a pathological mental illness. My parents, who knew no other framework for such occurrences other than the domain of psychiatry, asked for a mental health check from my doctor. It just so happened that my personal doctor was also my ex-GP trainer who had worked alongside me for a year and had certified me as being fit to be a GP only months before. Now he was being asked to asses my mental health.

My parents were also strongly motivated by the compunction to get me away from L, whom they saw as unsuitable and unpresentable to Bengali society as he was from a working class background. They were genuinely worried for my safety and the combination of my marriage breakdown, my 'unsuitable' affair and finding out that I was having visions, rang alarm bells for them as doctors.

Eventually, due to my parents' persistence, another ex-colleague and doctor at the practice where I was registered as a patient, turned up to the house I was living in ready to section me under the Mental Health Act. As is the legal protocol, she brought a psychiatrist with her. This particular GP knew me the least out of all the doctors in the practice, but was the most litigation conscious; she was also a part-time magistrate!

The psychiatrist who turned up at the house on that fateful day could see no evidence of psychosis, only the fact that I was extremely distressed. The GP signed me off work as sick and informed me that I would be referred to the General Medical Council (GMC) for an assessment of my mental health. Until then, I would not be allowed to work in case I put any patients at risk. It was known to everyone involved that no patient had ever lodged a complaint against my behavior, nor any colleague - this was happening purely as a result of my ex-husband's act of revenge.

You can imagine that I was devastated and destroyed. To have a GMC investigation hanging over you is what every doctor dreads in their life. And here I was facing an investigation when my medical career had only just begun. To know I was being hounded for my spirituality and also my expression of my sexual freedom as a woman by the very people who were supposed to be supportive of me was horrendous.

Yet at the same time I knew that I had created this entire scenario at some level of my being. I had tried everything I could to stop this from happening and it had not worked, therefore I knew that it was part of a divine plan, my own divine plan and I had no choice but to surrender to my own self (more on these principles later).

I had already realized in my own journey that there is no judgment in any situation. All situations are in perfect harmony; it is just our lopsided perceptions that think there is something is wrong. Once we adjust our perception to see the full picture we change our reactions to that situation. It is our own judgements that keep us in pain.

Interestingly, I didn't have these insights months later with the benefit of hindsight and some heavy-duty chanting, but made these realizations at the time that these events were unfolding. I immediately got a sense that, painful though the situation was, everything was perfect and that at some level I orchestrated it myself. I believe if I had not had this knowledge, I could not have handled the situation - I might have tried to commit suicide or really have developed a mental illness. But you are only given what you can cope with in life otherwise you would not be able to fulfill your destiny.

Love will Tear us Apart

A few days later, even in the middle of all my torment, I knew that L was actually still pining for the home he had left behind from before we had begun the relationship. I was also feeling a strong need to end the relationship with him and be on my own. I am so proud of what I did next as I could have clung to him in my darkest hour.

But even in my time of great crisis, I knew that the relationship was not going to work. Although myself and L are deep soul mates who have experienced many past lives together, it seemed that in this life, our time as lovers was over and we were not meant to go any further. It had come to a natural end; he moved out.

So here I was on my own with no job, no earnings, I had lost my house, my husband and everything that had gone before. The pain I woke up in every day was excruciating - my whole being would scream out. In order to try and alleviate the pain I would write for hours on my computer. As I wrote I would get answers from somewhere that would help me decide what to do.

I will never forget the pain that first Saturday after the break up with L. It felt like I had nothing left. I had to face a great big chasm in myself and my life. I had nothing to identify myself with and my old life was gone forever.

It was then and only then, when I had nothing else, that I finally turned to my passion in life - science. For almost ten years, I had had a secret hobby - the science of spirituality. There, that weekend, having lost all that I knew and in order to alleviate the terrible emptiness, I actually took my hobby seriously. I realized that this was all I had left. I still have the jottings that I did that day. They were eventually shaped into a book published four years later.

Jump

In order to recover from the shocking situation, I knew I needed to go back to my own house. My home was familiar and it was a healthy influence on me. It was by a canal and the flowing water of the lock gate was extremely healing for me. I started to visit my ex-husband, first on his birthday and then on further occasions. I did nothing but sit in his presence which, despite everything, was actually therapeutic for me. He was now mortified at the turn of events and my referral to the

GMC. He had never meant for things to go that far and promised to help me through the ordeal.

I knew in my heart that we could never get back together, but initially I thought there might be a hope. I so badly wanted to put everything back into Pandora's box, but there was no going back.

Staying in the house of the brother of the person you have just split up with when you have no money to pay rent is not a very good place to be. Their mother lived next door and the office of my ex-lover was in the back yard, accessible from both houses. I still felt very connected to this man and it was agony just to bump into him.

I needed to move. But without money to pay rent I was stuck. Somehow a life on state benefits just felt like defeat. I was sure at this stage that I would be reinstated by the GMC soon and I would be able to move wherever I wanted. In the meantime, my inner guidance was clear that I would have to move back to my own home to recover.

Strangely, I was getting asked to do talks all round the country and even abroad. People were constantly telling me that I was the best speaker they had ever heard. Just as many doors were opening as were shutting. Slowly, but surely I was getting into my Genius Groove.

Chapter 7

The Power of Goodbye

"It's not the game

It's how you play

and if I fall

I get up again" [h]

Madonna

In those first few months after I moved back to my home, I experienced the greatest agony I have ever known - it almost broke me. I suffered great heartbreak from the break-up of the relationship with my lover. I couldn't make sense of why we were so connected. One day, whilst journalling at my computer, I got the answer that we were 'twin flames'. I had never heard this term and dismissed it.

Then one day I got a call to be a speaker for a group who published a journal and they were also asking me to write an article. I flipped through the journal and found an article about twin flames! Here suddenly was proof that I had actually been communicating with a higher source - a part of me with infor-mation that I didn't consciously know.

I finally had confirmation that I was on a certain path, one that had been set up by me before I was born. I could no longer make things happen like I used to. I tried to get jobs, but that didn't work - my computer would crash or the phone line would go dead. My attempts to leave would be sabotaged. Difficult though it was living with my ex-husband, it seemed that I was being shown that I had to stay put.

Somehow I always had just enough money to pay my bills from various sources which included giving talks, my work as a healer and at the Bristol Cancer Help Center. I would plan ahead for several months for every long trip I made in my car so I could afford petrol.

A Town called Malice

All the while, the GMC were launching an investigation into my mental health despite the withdrawal of complaints by both my ex-husband and my parents. It seems the GMC have a vendetta against doctors who practice any form of complementary therapy. For example, Dr Michelle Langdon, a North London GP, was falsely reported in the tabloids as having used complementary therapies inappropriately.

As part of the GMC investigation, I saw several psychiatrists, two of whom said I was totally sane. One even asked me to come and give a talk to her colleagues! However, one psychiatrist had made up her mind before she had even met me. As I entered the room, she explained to me that I was ill. Nothing I could say could change her mind - it was already made up. After agonizing months of silence (the GMC are a really horrible organization) her verdict was that I had been ill, but was now better.

I had a follow up appointment with this same psychiatrist and took along over twenty-five letters of reference from various professionals gathered with just two days' notice, plus proof of all the activities I was now lecturing for a government scheme called Sure Start (via a consulting firm), presenting on the radio etc. In the previous appointment, she had assumed that these activities were not really taking place and were evidence of mania.

Her final conclusion was that she realized I was not manic, but the evidence that my medical husband and parents had submitted against me - details of all my visions and beliefs - were too strong for her to ignore. Although she and the other two psychiatrists could not find any evidence of mental illness in me, the reports handed in by J made a strong case in their eyes.

The Bitterest Pill

The GMC issued their verdict; I could go back to work, but only under supervision by a psychiatrist. They gave me a list of stipulations including that I was not to leave the country without telling them and, if the supervising psychiatrist saw fit, I must take prescribed medication. In addition, the whole world would be able to see I was being supervised on the GMC website.

All this had happened despite not having a single case of misconduct with any patient brought against me. No complaint had been lodged from anyone in my work situation. I was suffering the consequences of my ex-husband's vengeful and calculated attempt to destroy my career.

After a sleepless night, I knew what was being asked of me. I knew that I had to leave the medical profession and, in a one-lined letter to the GMC, I asked to be struck off the register. I was being asked to step into a greater reality, free from the restraints of the box that I was living in.

Later, I confronted my ex-husband and asked him why he had given out such graphic details of my personal life. He wasn't someone to ever talk about anything, especially difficult subjects. But when I realized the extent of what he had done, I asked him about it. His reply was chilling and the only time I ever heard him admit to it. He said, "I wanted to bring you down. I wanted to see you crumble."

The Power of Love

Still, I had a lot of work to do. I knew that to wallow in toxic resentment was not going to be the way forward. I spent hours at the computer wading through my pain. Trying to find out what it was *in me* that was creating the situation, I read Dr John Demartini's book, *The Breakthrough Experience* and felt a lot of release when I realized that my family and I have had an agreement at a soul level to behave in this way with each other.[1]

They had lovingly taken on the sacrifice of being temporarily hated by me, whom they loved and my ex-husband had agreed to lose me for ever - despite everything, he really did love me. When I realized that as souls, they were just taking on the mantle of being spiritually dumb to play their part in the game so that I left medicine and entered my true path, my heart opened with gratitude and humility.

In reality there is no spiritual and non-spiritual. We are all God playing out various parts and on the earth plane. People can appear to be ignorant, but they are truly infinite beings just taking on a disguise. When I started to live this truth, my path of recovery accelerated.

I knew that the only way forward was for me to understand that there is nothing to forgive. I started to do a lot of emotional work that consisted of self-examination to find the root of my fears; both in this life, in my childhood and in past lives. Why did I repeat the same patterns in life again and again? I started to understand, because life was showing me, how time works. I realized that every thing is actually pre-ordained and that a greater plan was laying out before me.

From the ego perspective, we don't know everything in life. But we are also great beings who have planned the most loving lessons for ourselves - even though they don't feel that way at the time. I was surrendering to my Higher Self. My Higher Self and my ego self started moving as one being and one voice. I stopped feeling the need to channel any more as I realized I am always channeling whenever I speak.

The more emotional work I did, the more I moved forward. I started to write the book, *Punk Science* and found that for the first time, I was not expected to go to work, as my family had been the cause of my losing my medical career. I could focus on recovering and writing my book. My ex-husband was just about able to to pay the entire household bills, so for the first time in my adult life, my time was free to think about what I wanted.

The more healing I did, the more my life moved forward as I resolved the issues that were holding me back. It was extraordinary, I would attract the next piece of the puzzle for my scientific work, by resolving my personal emotional issues. Gradually a scientific picture was emerging before me.

New Power Generation

I had been studying an exciting New Science for some years and felt that I had a part in it. The journey had started after I had the awakening experience in college. I became fascinated by the world of physics and how these cutting-edge ideas looked very much like esoteric wisdom. My aim at the beginning was just to produce a book that was easy to read, but in a genre that already existed - the science of consciousness and spirituality.

I travelled to New Mexico in 2001 and 2002 where I met some of the main people involved in the movement at a conference run by The Message Company.[2] They were effectively starting a revolution in science by placing consciousness at the heart of reality as a result of the logic of quantum physics. However, their books were often quite difficult to read for non-scientists. I noticed how little they incorporated each other's work into their own and how often the latest physics such as string theory was being simply ignored.

I felt that someone had to piece it all together and after a lot of resistance I realized that that someone was me! I started to really think about the writing a book in early 2002 after almost a decade of research, but I was hesitant to 'blow my cover' within the medical world.

It's amazing - when you start on your true path the universe really does conspire to help you. In my case, as I have just described, I then went on to lose everything else in my life and therefore had to follow my true calling and creativity.

In those early days, I gave talks simply emphasizing the parallels between science and spirituality, but would feel there was more to say - that there was more to discover. I read a lot about cosmology, the science of the universe as a whole which includes Big Bang theory. This world had undergone a revolution at the turn of the millennium and as a result was a very exciting field. I loved learning about dark matter, the cosmological constant, quintessence and other subjects.

The subject that most fascinated me was black holes. A lot of the data that modern telescopes were beaming back to us didn't fit the expected picture of black holes. I knew that there was something here that was very important. It was like I was remembering the future. Every time I did some emotional work I was given the next piece of the jigsaw puzzle.

Feels like Heaven

In September of 2003, I had written the first draft of the book. Then when walking through the woods one day with my dog, I paused to sit on a low branch of an oak tree. I started to tune in to the rotation of the planet, having seen a clip some years before of Brian Swimme the mathematical cosmologist lamenting how we never allow ourselves to do this.[3]

To my surprise I was suddenly in an aspect of universal consciousness that I had never seen before. Suddenly all the data about black holes just fell into place. I realized that I had just been given information essential for the next big discovery in science. When I got back home, I couldn't stop myself from falling asleep. When I woke up, I had the next part of the vision - it was given to me during my sleep.

For the next few years, I was given more and more proof of what I had seen. I wondered - why me? Why was I given such a huge vision for humanity? After a while the strength of the evidence was outweighing my reluctance. I realized this was much bigger than me and I had to take this to the world. Even though I didn't have a publisher for my book, I continued to write over the period of four years.

Sweet Surrender

It was as if I had surrendered all control of my life and now lived this new reality where, ironically, financially I was taken care of even without a job and I was only required to follow my heart and go where I needed to go. To move forward on my path I was either given a sign from 'without' or felt moved from 'within' to go in a certain direction.

What I quickly realized no longer worked in my world and still does not to this day, is to make a plan and try to execute it. Before this period of my life, I just assumed that this is the way the world works. But I found that trying to make something happen or even using the Law of Attraction did not work anymore. I had to realize that this was a plan that was unfolding perfectly and there was nothing I could do about it, much to the consternation of my ego which was still trying to control everything. This state of surrender eventually became second nature to me.

I headed down the path that was being laid out for me, having to surrender my ego desires for material wealth. The strange thing was that as I started down the path, my material needs were catered for until eventually I started living in luxury. Paradoxically, the more I surrendered the need for wealth, the more I had in my life, but instead of earning money through a job I hated, I was doing exactly want I wanted everyday. Now I have more material wealth in my life than I would have had working as a doctor.

Most of all, I knew I just had to keep writing the book, *Punk Science*. I had not written anything creative in the ten years since I left school and knew nothing about publishing, yet I persevered. It was so difficult to keep going when everybody was asking, 'what do you do?' If I said I was writing a book they would always ask if I had a publisher to which I had to reply, 'no' and watch their looks of derision afterwards.

Pretty soon, I realized these reactions were all judgments of myself and if I uncovered the underlying emotions of self-derision, I would not hear these comments anymore as they no longer reflected my own shame. It actually worked!

In fact, throughout all these years, this is how I moved my life forward - by uncovering my emotional shadow. My life had been flattened and all my insecurities were exposed in front of the world. I lost some friends who were disgusted at what they saw as my fall from grace as a respectable person and medical doctor to someone who no longer had status in the world. Worse still, in their eyes, I had voluntarily thrown my 'nice' life away.

There was no other option but to face myself and keep writing my book and continue exploring the dark side of my emotions that had been so violently flung open. These processes prepared me to receive the Black Hole Principle - the vision of the universe that I outlined in *Punk Science*. In the following chapter, I will summarize this theory and why it is so important and has huge implications, not only for science but also for our personal lives and for our Genius Groove.

Part Four - Lucky Star

Chapter 8

Waiting for a Star to Fall

"Think about it,

there must be higher love

Down in the heart

or hidden in the stars above" [i]

Steve Winwood

Shattered Dreams

A t the end of the last century, we thought we had it all sorted; we thought we knew all there was to know about the cosmos - how it started in a Big Bang and has been developing ever since. If you think about it, it's quite a strange concept that we could ever know about everything there is to know about this vast universe from our position on a seemingly obscure blue planet rotating around a medium sized star in an unremarkable galaxy. But it is true - we thought we had it all wrapped up.

Then in 1998, our concept of how the universe works went through a major overhaul. Soon our previous arrogance become all too clear as new evidence sent us back to the drawing board. Back then, I was a Senior House Officer in the Accident and Emergency department of Hemel Hempstead Hospital, Hertfordshire, UK. One morning, after a night shift, I was slumped in the staff coffee room summoning up the energy to go home, when the television displayed an image of galaxies receding away into the distance.

The breakfast news was reporting on the fact that new evidence from cosmology had revealed an unexpected picture about our universe. From the early days of the twentieth century, it has been widely accepted as scientific fact that the universe began with an enormous explosion, the Big Bang, and has been expanding ever since. However, it was believed that this expansion was slowing down due to the all the 'stuff' inside the universe; galaxies, stars and planets, me and you; pulling the universe back into a Big Crunch through a gravitational pull. So although the universe was believed to be expanding away from that initial primordial explosion, it was thought the rate of expansion was slowing down.

Not so it seemed - new observations of supernova indicated that the rate of expansion of the universe was speeding up, not slowing down.[1] The universe is running away from us! It is being blown apart by a mysterious force, powerful enough to overcome the gravitational pull of all the objects of the universe. This mysterious force was was named 'dark energy'. Sitting in the staff room that day, I acutely felt the stark difference between my job as a doctor and my true passion - to understand the secrets of the cosmos.

Keep Feeling Fascination

The turn of the century was an exciting time in cosmology and I lapped it up. To fuel my secret habit, I would sneak cosmology articles onto wards, stuffed into my white coat pocket and whenever I got a spare moment, which was not very often, I would read about the revolution. Perhaps I should have been studying how to fix a broken ankle, but that didn't seem as exciting as understanding the universe itself. Words like 'quintessence' and 'lambda constant' held much more of a fascination. I was especially interested in the work of maverick physicist, Joao Magueijo who said that the speed of light was not the ultimate speed limit of the universe, something my developed spiritual abilities found fairly obvious, but here was a mainstream scientist saying so.[2]

My secret studies continued and evolved as I trained as a GP. I discovered that the shock about dark energy and the runaway universe was not the only surprise in store for cosmologists. As the data from more modern powerful telescopes started to pour in, the emerging picture of the universe was far different from their cherished ideas about it.

What fascinated me the most was black holes. Like most people, I had heard about black holes in the context of science fiction movies or documentaries on TV. I thought of them as dark, exotic guzzling monsters, gorging on everything that slipped past their event horizon - the point of no return around a black hole. Even light could not escape the clutches of a black hole, hence being called 'black'. I didn't really know that much more than that.

But strange things started to occur that directed me to find out more. I was walking through a bookshop one day and a book fell off the shelf in front of me. It was written by Professor Stephen Hawking of *A Brief History of Time* fame along with mathematician Sir Roger Penrose.[3,4] I bought the book despite the fact that I could hardly understand a word, as it involved many equations and mathematical diagrams. I discovered that Stephen Hawking was quite pivotal in changing our views of black holes. Back in the 1970s he realized that if black holes do indeed guzzle everything around them, not only would all the objects that fell into a black hole disappear forever, but the information contained in the object would also be lost. This would cause an imbalance in the universe which troubled the young physicist.

He suddenly realized that if the black hole could radiate very weakly, then the problem would be solved.[5] Remember back in Chapter Five, we talked about the universe being filled with the vacuum energy or the Quantum Vacuum? Well this process occurs everywhere and near a black hole is no exception. According to what is now known as 'Hawking radiation' the process of the photon splitting to become the positron of antimatter and the electron of matter is occurring everywhere in the universe and even around black holes. What Hawking realized is that if one of the particles falls into the black hole and the other is radiated out, the balance of information of the universe would be restored. Hawking radiation changed the reputation of black holes so that they weren't seen as quite so black anymore, they radiated matter very weakly.[6]

It seemed that everywhere I looked, I kept finding references to Hawking radiation and black holes. However, I didn't know what to do with it - it felt like I was trying hard to remember something - like a dream, now vague and half forgotten. As you have read in the previous chapters about my life, my career and personal life had collapsed around me and I was left suddenly having the time to ponder the universe - a situation that felt divinely ordained. It felt as if I was to do something important that required the orchestration of such catastrophic events: as if some unseen plan was unfolding.

Electric Avenue

So I continued my studies and read what I could of various science journals. They often contained information that could not be explained by the current scientific paradigm. Page after page threw up strange findings that did not fit the received wisdom about the cosmos. In particular the strange data about microquasars captivated me.

Quasars are very bright objects in the sky that appear to be spinning very fast, so fast in fact that it is believed that they have a black hole associated with them. Microquasars appear to be the same type of objects except very much smaller. The behavior of a large quasar is thought to be the same as a small one, only at a different scale.[7]

We are told that anything near a black hole is sucked into oblivion. However, observations of these microquasars reveal that they are actually spewing out material at great speeds: almost the speed of light in fact. Astrophysicists did not understand this at all, but they initially observed electrons apparently traveling *faster* than the speed of light.[8] They explained away this anomalous finding as an optical illusion, but I found it intriguing. How and why do electrons reach such speeds?

But that was not all, microquasars also give out positrons and gamma rays. Gamma rays are caused by very high energy particles of light or photons. These particular gamma rays have the energetic signature of when a positron and electron come together to create light, as discussed in previous chapters about the Quantum Vacuum.

What exactly is going on? These and other puzzles were foremost in my mind in the autumn of 2003. As you can imagine finances were tight due to the loss of one of the household incomes. This led to my ex-husband demanding that I put on some courses in order to create some income. I agreed and soon, I had organized two courses in Berkhamsted in late September and early October.

The first course was not as well attended as the second with only three delegates - but what a group we made! We were all dedicated and immersed in science, which was such a pleasure for an enthusiast like me. The whole day involved both my lecturing as well as periods of group dialogue. The dialogue part was extremely exciting with such a knowledgeable group.

We were discussing the nature of gravity and then suddenly, one of the group said that he believed that gravity could be the 'friction' that light created as it moved through the dimensions of the universe. Something about that concept struck me as true. I suddenly knew that light created all known forces.

She's Got Herself A Universe

During the following week, I had the vision that I previously have described whilst sitting on a branch of a tree and musing on the Earth's rotation as described in one of Brian Swimme's videos. [9]

This was when suddenly, the universe opened up to me and revealed its secrets. It is hard to explain if you have never had a mystical experience, but I was *within* universal consciousness itself. I saw how all the data to do with black holes fitted together - how Hawking radiation *really* happens.

I got a sense of just how infinite the universe actually is; our current science is laughably inadequate to describe it. The human race will never truly know all of its complexities. In fact, I saw that the universe is both complex and simple at the same time - as if a few simple mathematical and geometric rules kicked off all of creation to become the world we see around us with all its rich variety.

These rules are similar to the algorithms that we use on computers to create fractal pictures such as the Mandelbrot set.[10] To create the fractal picture is relatively simple, but the pattern repeats infinitely creating a pattern within a pattern within a pattern. The universe as a whole is similar to that, only we don't understand the rules.

I also saw the truth about black holes unfolding in front of me almost like an animation. They are not the great guzzling monsters we have previously believed them to be, but they are the creative sources of the universe and everything we see around us. I saw the infinite source of everything at the center of the black hole, something we call the singularity. Light emerges from the singularity and steps down the dimensions until it reaches the boundary of our dimension, currently called the event horizon.

When light reaches this boundary it splits and becomes the positron and electron pair that we see coming out of black holes and objects such as quasars. But that is not all, sometimes the balance shifts the other way and they recombine to form light. We see this pattern of emissions coming out of black holes via telescopes: periodic flashes of gamma rays which have defied explanations by cosmologists.[11]

Chariots Of Fire

Gamma Ray bursts are very powerful and can reach us from the outer regions of space. They are so strong that they confound current physics thinking as they defy Einstein's theory of relativity - they have so much energy.[12] Scientists currently think they are caused by explosions - such is the power and violence of gamma ray bursts. However, observations do not fit the theory of explosions being the cause. Instead of one big explosion that fades away over time, gamma ray bursts can be repeatedly emitted by a source over many hours, days or even months.

Gamma Ray bursts are interspersed by X-rays, sometimes referred to as an 'afterglow'.[13] This name paints a picture of a high energy burst explosion which gradually fades away. Except this is not the pattern that is actually observed; the so-called 'afterglow' doesn't necessarily fit this picture. Often you can see gamma rays and X-rays following on from each other in a series of rapid bursts - a pattern that has actually led cosmologists to say that black holes 'burp' on their food.[14] The evidence simply does not add up to the picture that is presented by modern cosmologists who still believe that black holes are guzzling monsters and therefore should not be emitting material.

The vision that I had, made me realize that black holes are responsible for the creation of gamma ray bursts. Not due to a violent explosion, but due to the movement of the light spiraling out from black holes, shifting gently in balance one way or the other. When the light shifts in one direction, particles of matter and antimatter are produced. These are the positrons and electrons which we see coming out of black holes at the speed of light.[15]

As the balance tips the other way, the positron and electron combine together to form light which is released as a gamma ray burst. This shifting to and fro is a bit like breathing, or as the ancients called it, the 'flow of the tao'. Neither direction is more important than the other; what is important is the flow between the two. This process is the reason why we see strange patterns through our telescopes that reflect the process described above - the so-called 'gamma ray burst repeaters' and the apparent 'burping' of black holes.

Why are gamma ray bursts so powerful? They are the results of processes occurring in dimensions of the universe that we have, until now, totally ignored in science - the dimensions which lie beyond the Perception Horizon.

Chapter Nine

Velocity Girl

"Faster than the speeding light she's flying

Trying to remember where it all began" [j]

Madonna

The Land of Make Believe

It is particularly odd that cosmologists are claiming they know everything about the universe as so much of their so-called 'standard model of cosmology' seems to be full of mysteries. Cosmologists were rejoicing when the first data from microwave telescopes came back, because they believed it gave them the answers to what the universe is made of.[1]

They concluded that the seemingly solid matter we see around us and out in space - humans, stars, planets and galaxies makes up only a very small percentage of the universe - less than 1% in fact. What makes up the vast majority of the universe is a lot of stuff we simply do not know very much about - dark matter at around 23% and dark energy at 73%.[2] Bearing in mind that we didn't even know dark energy existed until a few years ago, it seems premature to proclaim that we understand the whole universe. (In this standard model picture the rest of the universe is made up of intergalactic gas.) So what and where is all of this stuff? I realized there was more to the universe than literally meets the eye.

Quicker than a Ray of Light

As I said earlier, I had been studying the work of a physicist called Joao Magueijo who had published theories on Varying Speed of Light (VSL)[3] This theory had been put forward to solve some of the issues with the Big Bang theory (namely the horizon and flatness problems).

Magueijo and his team developed the VSL idea saying that light was not always the speed that it is now - in the early universe it could have been faster. When they published their theory, there was a furore. The controversy stimulated by the self-proclaimed 'punk physicist' propelled the sale of his book, *Faster than the Speed of Light* to the best-seller list. In all the publicity, I found an interesting nugget. In a 2003 interview with journalist Robin Williams he revealed his true thoughts about light. He believes that light is actually infinite and is expressed through different dimensions of the universe.[4]

Suddenly I understood that the speed of light is not the speed limit of the universe. In fact, Einstein's famous law actually said that any object which has mass cannot go faster than the speed of light. Something without mass - eg light itself could go faster than the speed of light. So if there is lots of light in the universe beyond the light barrier, where is it?

In another flash of inspiration, that occurred this time in my kitchen, I had the answer - it lies beyond our normal perception. In other words the speed of light is not a speed limit at all, it is a reflection of the vibration of consciousness that is the limitation of our three-dimensional world. I have renamed the speed of light as the Perception Horizon.

To make this leap of logic you need to incorporate consciousness into your view of the universe which is normally ignored by science, but which some quantum physicists are saying is fundamental to the universe. Einstein's theories did indeed incorporate perception as he said that space and time behave differently depending on where you are in the universe and how fast you are traveling i.e. it is dependent on your point of view! So it is not such a giant step to incorporate consciousness into our universe. Nothing with mass can exist beyond the Perception Horizon, but light and consciousness do not have mass. So consciousness can exist in different areas of the universe that we did not know about before. Could it be possible to move your consciousness beyond the Perception Horizon? Is this the basis for the mystical experience?

<u>Dancing in the Dark</u>

The concept of there being other dimensions beyond the Perception Horizon of our dimension could also explain why dark matter and energy are so elusive. We have been looking at the universe in a very limited way, but there are aspects to the universe beyond normal everyday perception. Dark matter and energy could exist in these regions except now, we know that they are not dark at all, but light expressed in other dimensions.

The light above the Perception Horizon is not light visible to our normal consciousness - it is light of higher dimensions or 'superlight' to borrow a phrase coined by the engineer, John Millewski, who has also concluded that light is emitted from black holes.[5]

We normally think of black holes as the darkest places in the universe. In this new perspective, everything that is dark is actually superlight. Black holes then become the brightest place in the universe. Now we can understand why we see electrons and other particles coming out of them at the speed of light. Black holes have the source of infinite light at their center: the singularity. Light then makes the step down process through all the dimensions before reaching what is now known as the Perception Horizon. At this point it splits into an electron and positron and we see these particles appearing to come out of nowhere because we are not aware of the higher dimensional processes producing them, beyond our perception.

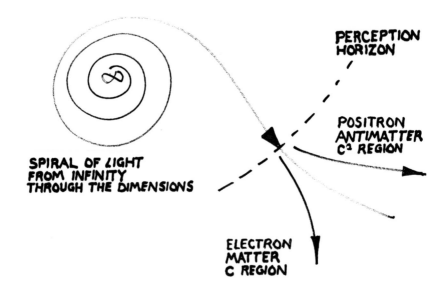

SPIRAL OF LIGHT
FROM INFINITY
THROUGH THE DIMENSIONS

PERCEPTION
HORIZON

POSITRON
ANTIMATTER
C^2 REGION

ELECTRON
MATTER
C REGION

That is why electrons appear mysteriously to be traveling at the speed of light. It is not because they are being whipped up by magnetic fields around the black hole - nobody has ever been able to explain how these fields manage to reach such powers. These electron jets are actually the last step in a process that occurs in an area of the universe unknown to us. The universe is far greater than we can imagine.

Heaven is a Place on Earth

My vision in the woods did not end at galactic black holes. As I left to go home that day, I felt an immediate need to fall asleep; I could not stay awake. When I woke up, I realized I had another piece of the jigsaw. For not only do black holes behave in this way, but atoms do also. My dreams showed me that the movement of electrons and positrons and gamma rays up and down the energy shells of an atom is actually exactly the same breathing process that occurs in galactic black holes.

Instantly I knew I had found a fundamental pattern of the entire universe. In the consequent months I found evidence that planets, stars, comets, galaxy clusters and even objects in our own solar system, including the Earth, show the same pattern. For example, terrestrial gamma ray flashes have been detected in the Earth's atmosphere by satellites and amazingly they have similar energies to the black holes in outer space! [6] I found that antimatter and matter and light in an interwoven spiral is emitted from the sunspots on the sun, antimatter fountains pour out the the center of the Milky Way, our galaxy shows signs of breathing movements - the evidence just kept building.[7]

In other words everything is a living, breathing black hole. The universe is a fractal of the same pattern. In fact, there are many physicists who are arguing for a fractal universe rather than the Big Bang theory.[8] Creation did not happen in one moment 15 billion years ago, it occurs all the time. Every moment at every level of the universe, infinite light travels through the dimensions and to the Perception Horizon where matter and antimatter are created.

Even our own bodies are created like this. This is the true science behind those much maligned chakras which are dismissed as New Age 'mumbo jumbo'. Chakras are energy centers in the body that cannot be detected when the body is dissected so our current science dismisses the concept as nonsense, despite the fact that the concept occurs in many cultures around the world and even appears in the Bible in the Book of Revelations. Some scientists have actually studied and made measurements of the chakras. [9] We are also made by the same processes as a black hole, every cell, every atom, every subatomic particle. In *Punk Science* I have listed a lot of the evidence for these ideas in detail.

Sky Fits Heaven

There was another issue - where does all the antimatter go that is produced in black holes? Antimatter is a thorn in the side of cosmologists. It is the exact counterpart of the matter that makes up our known world and there should be equal amounts in the universe, yet we can find hardly any of it. Cosmologists answer this by saying that shortly after the Big Bang, matter and antimatter had a great battle and somehow matter won.[10] When pressed to explain how and why exactly matter won this battle, only circular arguments exist, 'It won because it did'.

Yet in this new model of the universe called the Black Hole Principle (BHP), antimatter is being produced in droves. I realized, having read the work of William A. Tiller and others, that antimatter does not exist in our universe, which exists under the speed of light, but in a slightly faster moving world: the world of the speed of light squared (c^2).[11, 12] Strangely this is a world of negative time - time that moves backwards relative to our world. This concept of negative time is not as bizarre as it seems. It appears in some of the latest theories on cosmology like Loop Quantum Gravity.[13] It has also been linked to the world of antimatter before, with some authors linking it with c^2.

Suddenly all the pieces made sense. The part of a black hole beyond the Perception Horizon was the world that is out of space and time. It is only below the speed of light and in the region of c^2 that time actually exists. Therefore in the big picture, both timelines cancel out so that ultimately time doesn't exist.

The Black Hole Principle also gives a larger perspective to the Quantum Vacuum. The cycle between light, antimatter and matter in the Quantum Vacuum are actually fractal black holes operating at every level of the universe. I was at a conference held by the Institue of Physics when the speaker announced that Werner Heisenberg, one of the founders of quantum physics, originally believed that the universe was filled with infinite black holes, but did not trust his answer.[14] It is an interesting anecdote from history that the original inception of the Quantum Vacuum was as a sea of black holes!

<u>One Way Or Another</u>

In higher dimensions there is no time, but below the Perception Horizon of both the c and c^2 region, time exists. However, remember that the electron and positron regions are just light in disguise. They take on a temporary persona in which they appear to be separate.

Time exists and does not exist at the same time. The negative timeline of the c^2 region contains all future events. The c region contains all the events that are unfolding around us and is the world that we are used to. In the big picture both timelines cancel out so that in the higher perspective there is no time: only eternal timeless light. Yet below the Perception Horizon time also exists. Both states exist at the same time, there is no paradox. Since the dawn of quantum physics we have become more comfortable with something being in two places or many states at once.

The Black Hole Principle view of time is a radical discovery. Yet it resolves the paradox that exists throughout humanity's history - does time exist or not? Every mystic is trying to convince you that time does not exist. And that is absolutely true at a deep level, but they are also trying to get you to attend their seminar next Tuesday and will be annoyed if you turn up an hour late. How do we solve the paradox that we experience and witness time moving forward and are all hurtling towards aging and death in a seemingly irreversible journey and the concept of an Eternal Now?

Nobody I know of has been able to reverse their aging back to being a child through their consciousness alone; even the mystics that insists that time is illusion always die in the end. The picture is very similar in physics as relativity tells us that time does not exist. So how do we reconcile this with our everyday experiences?

Both are correct, time does not exist at the deeper level of the universe, but at the level of our everyday existence time does exist. We can experience a forward flow of time and, if we can become aware of it, the flow of backward flowing time as well - I call this 'living from c^2'. More on this topic later when we shall explore what BHP means for our lives and our sense of time.

You Spin Me Round

In the creation of everything from a galaxy, to our bodies or even our DNA, the light from infinity comes through higher dimensions to the edge of our reality - the Perception Horizon and then splits into antimatter and matter. The antimatter create the mirror world and our auras, just as a galaxy has a dark matter halo around it. (Some physicists believe that at least some dark matter is made up of antimatter.)[15] This antimatter region contains our subconscious emotions and the blueprint for how our lives will unfold.

At all times, part of our beings are also operating beyond the Perception Horizon through the dimensions and into infinity. We are now moving into an era in which we are becoming aware of our infinite nature and of the level of choice that we have - that we too are like gods and creators

As the Big Bang theory and the evidence for it crumbles around us, we are now faced with a new view of reality that creation is happening at every level in every moment.[16] It didn't just happen fifteen billion years ago. It is in the here and now - every single part of you is creating in every moment. Creation is a fundamental part of the universe and as evidence mounts for the fact that the Big Bang did not happen we can move into a new era of the continually creative universe.

Creativity is not just a nice buzzword coined by a management guru to get you motivated on a Monday morning; it is your cosmic birthright. You are an infinite being. The core of every part of you is a spiral from infinity connected to every other singularity throughout the cosmos. The part of you that is in higher dimensions and out of space and time smiles upon the part that is in the slower-than-light-speed world and creates loving lessons for the maximum growth on this plane, because this is the true richness of the soul's journey.

You Can Dance - For Inspiration

Having understood the Black Hole Principle and how the universe is in constant continuous creation at every infinite fractal level, it becomes clearer as to how fundamental creativity is to your nature. As we move away from the concept that creation occurred at a fixed point in the distant past in a Big Bang, we can also shift from the idea that everything has been discovered by clever people who are long gone and there is nothing significant left to create or understand.

Your creativity, your Genius Groove, is occurring in every single fractal black hole at every level of your being. Rev. Matthew Fox speaks of creativity as the place where the divine and the human meet and this is beautifully illustrated by the BHP and its balance between light, matter and antimatter.[17] It is through the dance at the Perception Horizon that we meet with the infinite and become our most creative selves.

Sometimes the balance shifts and we become weighed down in the matter and antimatter realms. We lose touch with our greater selves, dancing between this reality and the next. This can affect the freedom of our consciousness to really soar in creativity and our Genius Groove. What is it that can tip the balance? Our emotions!

In the next chapters we will explore how unresolved emotions can keep us held in space and time and prevent us from soaring in the multidimensions and how, if we release them, we can become freer to dance with the light of the divine, in our Genius Groove.

Part Five - Express Yourself

Chapter 10

What a feeling!

"I've been to paradise

But I've never been to me" [k]

Charlene

I am going to talk about one of the most ignored and the most misunderstood topics in the world, yet it is the key to The Genius Groove and this is the subject of emotions.

Human Nature

Whenever anybody mentions emotions, usually everyone recoils in terror. To be 'emotional' is often seen as an insult in Western culture implying you are feminine and not powerful in a male-dominated society. Boys are told to toughen up and not cry, lest they be like girls. Emotions are ignored, belittled and derided by many sectors of society.

There has been much written about how the scientific and religious paradigms in Western culture encourage suppression of people's emotions. However, there is another part of society that can be just as oppressive in its tirade against emotions, but is less recognized for it. I am talking about the spiritual and 'New Age' sectors of society.

So-called spiritual teachers have for literally thousands of years taught that emotions are a primitive aspect of life and that true enlightenment lies in rising above them, which can lead to their denial. These teachings have often originated in the East, but are now distributed around the world. People following these teachings can sometimes behave in a way that is incongruent with their beliefs, as their denied and repressed emotions are being played out.

In addition, there are many stories of abuse of power by some so-called enlightened gurus. This too can be a consequence of emotional denial. A recent example of such a case was the scandal involving gurus Eli and Gangaji who were originally from the USA, but trained in Indian traditions before developing their own international following. Eli and Gangaji are married, but in a shocking revelation to their followers Eli admitted to having had affairs with certain students in secret.[1]

I am not in judgment of his affairs, but pointing out that Eli's very own spiritual teaching points to the denial of worldly and ego desires.[2] He demonstrates someone who is preaching certain spiritual ideals, but cannot live up to his own teachings because he is not emotionally coherent. This is the danger with this type of spiritual teaching; a constant denial of certain aspects of life can lead to an emotional explosion.

Keep It Together

A lot of gurus teach people to rise above emotions, treating them as a human failing that has nothing to do with enlightenment. Guess what happens when you try and ignore something? Just like when you try and push a beach ball under the water, it bobs up even stronger. So people who are on the so-called spiritual path who try and ignore their emotions, attempting to stay in a state of bliss, divorced from everyday life, will find that their emotions start screaming out for attention in some form or another. This is why the enlightened master will suddenly break out into a fit of repressed emotions, be it rage, desire or a lust for cream cakes, due to the years of denial.

Elizabeth Gilbert in her best-selling book, *Eat, Pray, Love* describes her experiences of traveling to various countries after her divorce, including a trip to an Indian ashram where she is taught to meditate.[3] She sits down quietly and soon many of her emotions come up to be recognized. But she is supposed to be meditating and feeling calm! Emotional turmoil is not on the agenda. So she tries to suppress what is naturally bubbling up in her consciousness, which results in some very funny passages in the book as she enters into a battle with herself in the meditation room.

She has been taught, like so many others have, that emotions are bad and avoidance of them is good. She believes that any naturally occurring feelings that arise when the mind is quiet need to be pushed down at once. This is the system that has been handed down to us from Indian traditions as well as being taught by modern New Agers around the world. They teach that emotions are dirty and to be avoided. Love, light and nice cuddly feelings are all people want to deal with. All this leads to some bad cases of 'New Age Rage' whereby long repressed emotions start to explode.

Island Of Lost Souls

In my treatment clinic, I found that many of the people who espouse love and light have actually had deep traumas in their childhoods. They are usually women who have been sexually abused or had severe beatings as a child. If we go back to the concepts of Chapter Four - that everything is consciousness, then we can also describe ourselves as consisting of consciousness and existing in many dimensions, as consistent with modern physics.

The traumas in the lives of some of these women cause the 'consciousness body' to avoid dwelling in the lower part of their physical body. This is the site of the lower chakras and memories to do with abuse tend to get lodged there, because the lower chakras are associated with basic survival, safety and sexuality. As a healer, I am trained to detect the subtle, nonphysical field around a person that is sometimes called the 'energy body'. In some cases, the energy body becomes focused in higher chakras which means the person can avoid some of their pain caused by caused by unresolved emotions from abuse; their energy body blocks it out.

Because the focus of their consciousness is in their high chakras, they can develop extraordinary gifts of psychic ability. They also tend to want to live in a world that is comfortable: full of angels and fairies and nice things. This allows them total avoidance of the pain that they have experienced in this life. I have witnessed the total transformation of such a person as they suddenly exploded with the unresolved emotions caused by the memory of child abuse. Quite an amazing experience and a revelation of a different side to that person that is not as fluffy, but definitely more authentic.

So whilst a lot of people in general are oblivious to their emotions, the so-called 'spiritual' crowd are also suppressing their painful feelings with 'love and light' candyfloss. As my friend Jenna points out, we often use the term 'Mind, Body, Spirit' which is devoid of any mention of emotions.

Why is this so important? Because when you face your emotions and understand *all* of you, including the sides you do not want to face, because you think there will be something dark and dirty lurking there, you bring light and balance to these emotions and you become authentic. By resolving long-standing emotional issues, you liberate your soul and expand yourself into the divinity of the universe. In finding authenticity and your true self, your fundamental creativity is more easily revealed. In short, emotions are the key for you to get into your Genius Groove.

Keep It All In

I have just discussed how emotions are being ignored and pushed aside, but what do I actually mean by emotions? Well this is where there are further misunderstandings. Often when we think of emotions, we think of emotional extreme points such as anger, sadness, depression, joy. We think of our emotions as the just the visible expression of how we feel. These visible expressions can be seen as embarrassing, so they are sometimes repressed.

As young children we (usually) express our emotions at all times not caring what our societal surroundings are. After many years of having been told to stop expressing our emotions due to our parents' embarrassment, we shift from our carefree expressions of pure frustration or happiness to learning that the expression of emotions is not a good thing.

Having been brought up in England, but not in an English family, I am in between two cultures. I am always amazed how seldom emotions are expressed in many English households, whereas whilst I grew up, my mother would be crying one second then laughing the next, she would express how she felt in the moment.

I have come to realize that this is not normal in England. But it is not just the English who have perfected the art of emotional repression; it exists in many countries, cultures and religions. Raw emotional expression is rarely explored. Combine this with the active suppression of emotions from the New Age crowd who have deemed them not a worthy part of a spiritual life and you have the situation we are in now. Emotions are not deemed worthy of examination by society; it is recognized that we each have an emotional life, but most people leave it at that.

I Know This Much Is True

From my experience as a doctor, a bio-energy therapist, a scientific explorer and a mystic, I have learnt to see emotions as fundamental to life. In fact, I believe they are actually part of the very fabric of reality. This is not exactly a new idea, many esoteric teachings speak of us having an emotional body as well as our physical body. This emotional body is seen to be an overlay of the physical one. The body is also seen to have other aspects such as an astral and spiritual. They are like different dimensions of one person.

This view of reality places emotions in their own separate space and, in a way, has helped to foster their neglect. What I have found over the years is that emotions are an important expression of consciousness that occurs below the Perception Horizon discussed in Chapter Nine. We shall discuss the physics in more detail later on.

As consciousness is fundamental to reality, it is also fundamental to our own selves. It is the intelligence that runs deep to our cells and our very atoms, it is the reason why our fundamental molecules such as DNA behave in such an intelligent fashion. As we have seen, some physicists conclude that one of the consequences of quantum theory is that matter emerges out of consciousness and not the other way round.

The word consciousness is a very contentious word and means many different things to different people. People trained in the Vedic traditions have very definite ideas about what it means, as it is laid out in their teachings. However, I would like to discuss consciousness as pertaining to the quality of sentience and intelligence and sometimes awareness. When I refer to awareness, I usually mean when something is able to be articulated either verbally or at a feeling level. Sometimes I speak of awareness at a universe level, in which case, you can say that the universe is becoming aware of itself.

The very fabric of our beings has sentience. It is from this intelligent blueprint that our physical bodies arise. What I have realized is that emotions are a part of the very fabric of who we are. You might say emotions are aspects of consciousness that have *charge*. Everyone knows what it is like to feel charged with emotions, sometimes we can feel annoyed and aggrieved and that emotion will rush through us and almost dominate our lives. Sometimes we can feel happy and elated and this can also make us feel a rush and run our lives for a while. Some of us may have also felt a glimpse of what it is to feel centered and at peace for a while. What happens in your emotional dynamics affects your whole life.

They affect what you attract into your life, they affect the way you behave, they affect the way your body behaves and whether or not you get disease. Emotions are a fundamental force in shaping your life and essential to understanding who you are, but most of the time we ignore them completely. I have come to the conclusion that in order to make fundamental changes to your life, it is essential to understand emotional dynamics in your life and not ignore them. In fact this is the true liberating process of life, the true alchemy and the path to your Genius Groove.

Now, this is not always an easy path, it involves facing your shadow self. But after a while, plunging into the deepest, darkest side of yourself becomes a habit and you realize just what the real 'secret' is; the more you resolve your emotional dynamics, the more you become your authentic self and the more magically your life unfolds and therefore the payoff for all this work is to enter into a new world that is not always easy, but is filled with so much more magnificence and deeper understanding than before. You actually start to slip through the cracks of everyday mundane reality into the deeper and more authentic world of your true self.

Emotional Highway

Wow! Now how do you go about doing this? I actually discovered these universal principles of emotional resolution as a bio-energy therapist. When I trained in the Plexus bio-energy system, I hadn't been taught particularly how to deal with people's emotions, but if you remember, I was on my own healing journey due to the years of violent abuse from my sister. I have my sister to thank for putting me on the road to emotional self-discovery. It stimulated me to read self-help books and discover aspects of myself that contained unresolved issues.

So when I started seeing clients using Plexus Bio-energy I was already working on my own emotions and had made some discoveries as to how emotional dynamics play out. When working with clients with bio-energy, I found that the therapy would sometimes bring emotions to the surface. Sometimes, with my clairvoyant senses, I would find myself tuning into an emotional story from past lives or from the client's childhood.

Often, I would end up in some sort of dialogue in which suddenly a new perspective would emerge and the client realized that everything in their life was perfect and was meant to have unfolded in that way. (I say 'find myself' and 'end up' because my actions were not made from a normal conscious place, but from a place of total flow.) Sometimes the emotional resolutions were so powerful that within a few minutes, both myself and the client would have completely forgotten what the issue was.

I started to find that once emotional resolution had been reached, my consciousness seemed to enter into different dimensions where I could see the person's life from a higher perspective. This could take many forms; sometimes the past life personality would actually manifest in the room and interact with me, one time I turned into a Hindu goddess with many arms. Every single situation is different and it always fills me with awe and wonder at just how creative the universe is. I understand that these events can seem bizarre to some people, but experiences such as these are becoming more common.

Oh Father

What I realized is that most people have got therapy all wrong. Either they are trying to avoid emotions and rise above them, which is quite frankly the fast track to illness because you are denying a part of yourself. Or they are trying to relieve the emotional hurt and apportion blame or even forgive the person whom they believe has wronged them.

But often this doesn't lead to any emotional resolution at all. Forgiveness, something so often preached by religious and spiritual teachers, simply does not work. If you are very angry about something, such as a guardian figure having abused you as a vulnerable child, forgiveness feels as if you are saying their actions are OK and that you are letting them off the hook.

So many people fall into the forgiveness trap. They seem to be all pious and calm on the surface, but inside they are seething. They won't admit it for fear that they will not be seen as a 'good' person. They go through years of unresolved anger because this is what they feel they are supposed to do - we are taught that good people forgive others.

Open Your Heart

The real way to resolve any emotional situation that has a hold over you, is to see it from a higher perspective, because in the higher dimensions there is no judgement. It is important to emphasis this is not an intellectual process, but an actual shift in vibration.

Have you ever been through a situation which at the time was not a happy one, but with hindsight you got an 'aha' moment and realized that without the distressing situation, the good stuff could not have come your way? You felt liberated didn't you? You probably also gained a glimpse into the divine order and perfection that exists in the universe.

To really heal emotionally you need to see *all* of life in this way. Everything, no matter how stinky it seems, is actually balanced out - but sometimes you just don't see it from the perspective you are looking at it from. Although the balance exists in that moment, sometimes it takes the long term view to see things more clearly. This can mean looking to past lives and higher dimensions and it is by taking these perspectives that we can understand some of the more difficult emotional dramas. For example with murder, where is the 'aha' moment and balance in that?

I reached a new perspective on this issue in an insight that came to me one day, just as I was setting off for a meditation meeting. It felt like it was downloaded in just one 'hit'. I realized that nothing that anyone can do can be judged by anyone else. It is all part of the vast learning experience of the universal consciousness that we are all part of.

This only works if you know that we undergo many lives and that death is not really death, but a transition to another consciousness. In the big picture there is no judgment in murder in the long run, as there is no death - just learning. The murdered and murderer are both aspects of the one underlying consciousness, call it the Mind of God if you will, and both souls have chosen to experience the murder scenario. In the long run, in the history of both these souls, it all balances out and there is no judgment if you view the situation from a higher perspective.

This is the place I was taking people to in my clinics, where they could see the higher perspective about something that had happened to them. Why don't we all murder? Well it is just not what most of us want to do because we do not need to have that experience. But it is important to realize that each of us has performed every type of heinous act - in our minds as a potential act, in past lives or simply because we are all part of the universal whole. So we cannot judge anyone else's acts as they are our acts too.

Excitedly, I relayed these discoveries to my meditation group. However, they were not impressed at the thought of there being no judgement for deeds such as murder. This is just a step too far for most people. Interestingly, as I changed my perspective in this way, the people in the meditation group started to fall away from my life which brings me onto the next chapter and the principle of reflection.

Chapter 11

Second that Emotion

"Nothing takes the past away

Like the future

Nothing makes the darkness go

Like the light" [1]

Madonna

Knowing me, Knowing you

This is one of the most difficult things to face for anyone, but as my friend, Peter used to say to me - the Universe Reflects! On one sunny day he took me to the park and we sat down by a tree. He told me to look around. He said everything you see around you is in some way reflecting you and who you are: your state of mind and your emotions. I looked around me and saw a crow and then a few kids playing and couldn't really see the significance.

However, the thought stayed with me. A few months later I was harassed and bullied by a consultant. She was a lady who had worked her way to the top from quite humble beginnings with a big boost to her career when she married well. Somewhere along the way she seemed to have become extremely unhappy as she had ballooned in weight and was living apart from her husband, as well as being an apparent workaholic.

Her style of practicing medicine was the most litigation vigilant I have ever come across, which meant the hours were long and arduous for her junior staff. From the start, she seemed to have a problem with me. She picked up on my insecurities about my career, having failed exams in medical school and just frankly never feeling comfortable with the job. For much of my medical life I managed to muddle along and sometimes some unusual abilities would kick in; I sometimes had an extraordinary knack of diagnosing patients quite quickly. At times, I knew what to do without really knowing why.

This doctor picked up on my weaknesses, she was also known to be racist and started to give me a hard time saying that I would not be able to cope with her job. She even looked into my past and found that I had taken a week off work when my mother-in-law was ill. She also demanded to know why I had not performed well at medical school and I reluctantly told her about the troubled times I had there, which she did not keep confidential, much to my chagrin.

But she also gave me one of the greatest gifts I have ever received. When she was admonishing me in her office one day, she used a few phrases that were the exact phrases that I commonly used about myself at the time. They were words that were born out of my low self-esteem. Interestingly, I cannot remember what these words were because the issue seems to have healed. I sat in her office and the penny dropped. 'Oh my God', I thought, 'she is reflecting my insecurities. The universe does reflect - she is reflecting me.'

The next time things flared up and I had a run in with her, I took a deep breath and told her that I saw how difficult things were for her and of course I would love to have the opportunity to work for her - in other words I realized the game I was playing with myself and treated her with unconditional love and compassion from a centered place - much to the utter surprise of her secretary who had been called in to the room to witness our discussion.

This change in me so fully resolved issues between myself and the consultant that we left on good terms after I had worked hard on her team - I even gave her a gift when I left her team and got a lovely letter back from her.

She was one of my most powerful lessons in emotional reflection. When you are really triggered by something or someone - you can bet that you are looking at a mirror of some aspect of yourself that you are in denial about and repressing.

Finer Feelings

Some people reading this may have already spotted some flaws in this argument. What about child abuse? Are these children being reflected by their tormenters? Obviously not, but in order to solve this riddle you need to step into the higher dimensions once again.

When you view life from this perspective, you can see that everyone can be seen as guilty for everything. If you take things to an extreme it is because ultimately we are all part of the whole. We incarnate many times in order to have many different experiences. In one life you might be the most terrible, ruthless person, in another you may be very gentle and have things done to you by ruthless people.

The thing is that if you look at the entirety of all our lifetimes, they all balance out. Some people might be saying at this point, 'So babies should just put up with being abused - they deserve it?' Actually, in my experience it does not work like this. It is not so much the deed that is being balanced, but consciousness itself and it is more complex and wonderful than the simple, 'an eye for an eye and tooth for a tooth' mentality.

I had the most extraordinary client who taught me a thing or two. She and her siblings had been sexually abused as children by their father who strangely enough (or not) had developed penile cancer. The memories were just starting to surface in her, which is why she came to see me. On the first session I just thought, 'What am I going to do? This woman has been through something so traumatic and I don't feel I can help her'.

But I did treat her and slowly but surely she made incredible progress. Her career was kick-started and she looked fitter and healthier by the week. For her last session we were both almost a bit puzzled as to why she came, but we trusted in her decision, knowing it was for a reason.

Cherish

Actually, this was the most amazing session where it all came to together and I learnt an important spiritual truth. As I treated her, the emotions took us through to a higher dimensional perspective of consciousness and that's where I got the surprise. This lady was starting a career in baby massage, which is a very loving way in which to touch babies - not what she had experienced in her own life. From a higher dimensional perspective she held a transformational key that the world needed at this particular time of planetary growth and change.

She had been abused as a child, but instead of abusing children herself, she taught people how to use a loving touch with children! From a higher dimensional perspective, it is for this that the whole drama had been created. She is an aspect of consciousness that glows throughout the dimensions beyond her physical form as she is transforming the abusive touch she experienced a child into a loving touch for others. She holds it all in her being. Her acting as a transformative energy is essential for the planet right now as we move into a new era. We are all of us holding energy in our own particular way that is just perfect for what the universe and our planet needs right now. All of us are playing our part.

So there were lives that she set up when she was either the abuser or had the guilt of abuse. Remember that it all works on consciousness so the actual act isn't necessary, it is about the emotions. If beating your child is acceptable in your society then you may not feel guilty about it, which will not set up any emotional charge for you to act as a vibrational attractor for events.

Something in this lady's consciousness attracted the situations in her life, as is the case with all of us, which led her to be the transformative light that she is on the planet right now. One has to take the situation through to a higher dimension in order to understand the full picture and every situation is different.

So yes, we always reflect at some level, but it doesn't have to be from this life and it may not be a direct action but rather a judgment or even a guilt that one *might* do that action. In any of these cases, the action must be out of keeping with one's own conscience even though it might be acceptable to many people. We also might have taken on a transformational role about an aspect of life, like my client did.

I have also found many times that when I have just healed an emotion, I have felt a real alchemical shift happening in my energy body. It is at that moment that someone may come into my life that reflects the emotion that is just 'leaving'. It is as if it is more prominent as an attractor on the surface of your aura before leaving. When you observe this happening in your life, it can be quite amazing to see your past pattern so clearly in someone else. So the principle of reflection is not as straight-forward as it first seems; it can occur in subtle ways that require us to feel more deeply into what is really happening.

Loving The Alien

Having looked at the thorny issue of child abuse we can have a look at how the principle of reflection works in other areas of life. When we take on board that the universe reflects us, we start to understand that when we are really emotionally stirred up, it is because we are seeing a disowned part of us that we have not yet learnt to love.

Where this principle is most intense is in our personal love relationships and with our children. Have you ever come across or been that person who is attracted to someone that is so wrong, but for some reason there is this explosive attraction and you cannot keep away from each other? Or maybe you have fallen in love and at first everything is so wonderful and then you commit to each other and then suddenly all this ugly emotional stuff comes up and you are left wondering what happened to those rosy, hazy days?

Guess what! This is the whole point. Your partner is the one who represents those aspects of life that you are not able to face or understand about yourself. You manifest them in someone in your life in order to love these suppressed parts of you. This is why people who seem so wrong for each other can't be sepa-rated. They have a contract with each other to love the re-pressed parts of themselves that the other represents - a polite handshake on the street is not enough, they need to be in a situation where real love and really deep emotions are being thrown around. It is only in these circumstances, when we are deeply touched/disturbed, that we really have to face our suppressed selves. But this is also where we find our true liberation, by facing and loving the aspects of ourselves that we hide from the light, we become whole and this aspect of our shadow no longer leaps out when we least expect it.

When people fall in love and have children, the children will also express the couple's or family's collective repressed emotions in order for light and unconditional love to be shone on them. A family came into my surgery once who were 'very nice' and upper middle class, displaying polite English behavior. But their child was a teenage girl out of control who spent time in behavioral centers.

She was representing all the repressed emotions in the rest of the family and they had no choice but to love her, because she was their daughter and, therefore, they had to love a repressed part of themselves. From having met the family, I would guess that she represented the wild and angry side that the rest of the family had locked up. They were so busy being good, genteel people that their anger had nowhere to go but be born in her.

Push the Button

It isn't just families, any situation or any person that makes you have an emotional reaction is reflecting something for you. There are some people in the New Age scene who just use this principle to say that if something is bothering you that you are displaying that exact same trait.

In my experience, this doesn't always hit the button and heal the issue. Sometimes you have to feel for yourself what that situation is reflecting for you. You may even be showing yourself a place of self-judgment and are worried that you are displaying that trait? Sometimes it is not obvious, but put your attention there and you will get the answer bubbling up from your subconscious.

Sometimes we are triggered by what we see as an injustice in the world. If it is something that really annoys us, you can bet it is reflecting a repressed part of ourselves which could even be from past lives. Ever since I was a child and watched my mother do more housework and chores than my father, despite the fact they both had the same job, I realized that women get a bum deal in life. I went onto read more about the inequalities in society and by my early twenties I was an active feminist.

Imagine my surprise when a few years later, my friend gave me a deep tissue massage and, through visions and intuition, we both uncovered that I was a slave trader in a past life - a privileged white male who had abused his captive women! No wonder I had such a feeling of injustice!

Again the principle of reflection is not as simple as straightforward payback for bad deeds in past lives. The higher dimensions do not behave like this one. You could say that the currency is not the actual deeds but an interplay of consciousness.

To illustrate this I will recount a time when I had a past life remembrance in which I was both the murderer and the murdered. This was whilst someone was actually demonstrating a healing technique in front of a group and using me as an example. Both myself and the instructor demonstrating on me had quite a deep experience. We both had the taste of blood in our mouths as we relived the scene. At times I felt the consciousness of both parties. So it is not as simple as one life follows another and we payback for all our deeds, although our linear minds want things to be so.

I have been critical of certain New Age factions, but not all in the personal development arena choose to ignore the emotional aspects of life. I am delighted to have found teachers like Debbie Ford who has written extensively in a very straight-talking manner about facing the shadow self.[1] Carolyn Myss is another teacher who doesn't pull any punches when it comes to healing emotions.[2]

Lucia Nella has developed some interesting and unique insights and principles, a veritable science of emotional healing, which she presents in a simple, direct way.[3] Importantly, all these authors share their own stories - they are humble enough to reveal their own shadow to the world and their learnings.

Opposites Attract

Another brilliant speaker and author is Dr John Demartini whom I first saw in 2002 giving a lecture in New Mexico.[4] He not only confirmed the principles of non-judgment as being the only way to free yourself from your emotions holding you back, but also took these concepts to a whole new level by applying the science of quantum physics to emotions in a very elegant way.

If you recall, we discussed in Chapter Five how, light, antimatter and matter are always in a continuous cycle of annihilation and creation. You now know that this is actually an aspect of the Black Hole Principle but let's leave that aside for now.

Demartini speaks of how physicists see matter as actually being frozen light. That is if you break down the atoms that make up the whole universe, you find that photons make up the subatomic particles which in turn are the building blocks of all the elements of the universe. Therefore, photons of light are actually fundamental to the molecules and atoms that create everything we see around us, including ourselves.

As we have previously discussed, when a photon splits it becomes complementary particles of antimatter and matter - a positron and electron. Demartini explains that when the photon exists in the photon state it does not have mass and exists without space, time or charge. However when it splits into the complementary particles it gains mass, charge, space and time. It therefore enters the world we can measure with our senses, what Demartini calls the 'conditional state'.

He sees light, before it has split into the two particles, as being in the unconditional love state. In the unconditional state the photon has no space, time, mass and charge. In the conditional state, particles have qualities of space, time, mass and charge. If we remember, the photon, positron and electron are in total flux all the time. In a way they are always separate and together at the same time.

Because physicists including John Wheeler relate light to consciousness, Demartini goes on to say that consciousness, like light, can also exist in a conditional and unconditional state. In the *unconditional* state, akin to the photon, the mind is out of charge, space, time and mass - therefore gravity. In the *conditional state*, i.e. when the photon has split off into two particles, the mind exists as two polar particles. However, the two particles are really always part of the whole and when they come together, they annihilate to become the photon again, something I have described in previous chapters.

Demartini relates all of our experience to this process. We see our lives in polarity - as just one half of the equation. In actuality, all life is always in balance just like the electron is always just one half of the photon all along. But we don't always see this balance.

For example, if we are distressed by a situation, we don't see and disown the other part of a situation which would bring us back to balance. We are happy to discuss how something has hurt us. This keeps us in the polarity state of charge, mass and gravity; we get weighed down by our emotions. When this happens, our emotions have a hold over us and can be in charge of our actions and thoughts. So when something has hurt us or bothers us, we can have recurring thoughts about it. In a way part of us remain trapped there in that particular space and time

However, when we shift our awareness, we can see the whole picture and realize that the situation was healed all along and always in balance. As we shift our awareness to incorporate the healed version of events, the polarities then cancel out and the situation is moved into light. It then loses its mass, gravity, charge, space and time and sometimes we can't even remember what the situation was all about just a few minutes later. The part of our being that was trapped in the situation is then recovered - shamans call this soul retrieval.

Freedom Comes When You Learn To Let Go

Listening to Dr John Demartini in that lecture in 2002, I was amazed; he had discovered the science behind my discoveries about how emotions work. The key to forgiveness is not to try and let a person off the hook, but realize that there is nothing to forgive: that your life is unfolding perfectly.

When something has a hold on you emotionally, it is because you have disowned a part of reality and you need to shift awareness to realize that this situation is whole and always has been. It sounds like a radical statement, but Demartini has worked with this system over many years and has even helped people with severe situations, such as a woman who was attacked and gang-raped. He helped her to reach a place in herself where she was filled with love for her attackers and was even grateful that the incident had come into her life as she had experienced so much grace because of it.[5]

Whatever healing system is employed - true resolution can only happen when this process of collapsing the polarities occurs. When the polarity of a situation is resolved, it become light and you can reach a state of grace. Although this sounds nice, we don't remain in a state of bliss. The next emotional situation arises and then the next one. Life gives you more situations from which you need to learn.

Every time you are able to shift your awareness to heal the polarities in situations in your life, you move into a different zone of reality of unconditional love and grace. If you continually repeat this process, your life doesn't exactly become easier, but you gain a wider perspective. Life becomes much richer and more magical; it incorporates more. You can even start to have multidimensional experiences. As you become more authentic, you throw off the emotional charge and gravity weighing you down and can dance at the Perception Horizon with your infinite self.

In my own life I have discovered that the more you do this process, the more you move into a state of grace and unconditional love and this becomes so strong that you can overcome almost anything that is thrown at you - not through denial, but through integration and understanding: through true emotional alchemy. A lot of so-called spiritual people get puzzled once they have experienced the state of grace, wondering why situations still happen that bother them. So they go into a state of denial as their self-image is as an 'enlightened person' who doesn't get bothered by all these things any more. This is when emotional situations can become so explosive. Integration, not denial of emotions is key.

Although the understanding of emotional dynamics is important, there is no substitute for actually experiencing the alchemical shift yourself. For this, a teacher or an exercise from a book can be helpful. Dr John Demartini also did an exercise where he took us through a process that made us own our disowned reflections that we perceived only exist in other people. I will briefly explain part of the process here.

We made a list of 'negative' traits that we see in another person. We then wrote down a list of people who see us with those same traits we dislike in others, then went on to list people who have benefitted from us having these traits etc. To see the full process please read one of Demartini's books or attend his seminars.[6] Debbie Ford's books also describe exercises to integrate your shadow.[7]

Through these types of processes, situations become equilibrated. This is not the same as denial of the emotions. It is about moving into a state of realization that everything is balanced and in perfect equilibrium and always has been. These are the exact findings I was making in my own healing clinic and in my personal emotional journey.

These philosophies have been severely tested in my own life. As I described previously, it was knowledge of these principles that helped me get through the days of the GMC investigation and eventually heal the relationship with my family and ex-husband.

Nowadays, people find it hard to believe that I went through such an ordeal, they think I look so young and carefree. I believe it is because I regularly process my emotions and don't deny them, that I seem so light. I am not heavily weighed down by the gravity of the past although of course, I still have emotional charge like everyone else, otherwise I would not still be in this dimension!

Bring It All Back To You

We do indeed tend to get weighed down by our emotions, sometimes even reflecting in our physical weight. Usually nobody teaches us how to do our emotional housework and how to become aware of our emotional polarities and turn them into light. Most of us live in an emotional prison, but if we were to be given the tools, we could free ourselves.

Eventually, through repeating the process of turning emotional polarities into light, it becomes a habit and we learn the deeper reality: that life is a permanent state of grace but it is often in disguise. When this knowledge is firmly rooted in our beings, we start to live a different life, one not divorced from emotional situations, but with a different perspective on why these situations exist.

By continuously processing our emotions like this, we can experience the multidimensions, because as our consciousness is freed from polarity, it can soar above and transcend and for a few moments enter into the invisible realms beyond space and time, beyond this reality and beyond the Perception Horizon. By doing what I call 'the work', our being simply expands. We enter into our multidimensional heritage. This is the path we must take to throw off the shackles of the paradigm we are currently in and enter into a new world.

The old world is crumbling around us. Through our emotions we can get more in touch with authentic selves and enter the New Paradigm with a new reality. I received a channeling once that told me that we are like crystals and by doing our emotional work we are polishing our crystals which lets in more light. It is through this process that the Earth is moving into a new era. People often look to the outside for something to heal us and save us for example the common belief in the New Age that space ships are about to arrive and take all the good people off the planet.

Whether this imminent or not, the fact is that the planet is shifting through each and every one of us. We all make a difference to this shift, because anything we do, or even think and feel, impacts the holographic field. It is through this field that information in consciousness is made available throughout the cosmos.

Holding Out For A Hero

Nowadays when I meet someone passionate about something or someone really driven and successful, I ask myself - who hurt them? What emotional drive ended up pushing them to succeed beyond everyone else? Or if someone is really active for a cause - saving children perhaps, when I get to know them, I invariably find out that when they were a child, they desperately wanted to be saved. This is all part of the non-judgement and balancing principles of emotions.

Let's examine the life of the most famous and successful performer of our times, Madonna, who also features in this book a few times. She lost her mother to breast cancer when she was very young and it seems that the resulting emotions may have driven her to being the top performer in the world. If you have any doubt that Madonna has been 'driven by her demons', you only have to listen to her lyrics as they are revealed in songs such as *Mer Girl* on the *Ray of Light* album.[8]

The heroes of our age are often quoted as Mother Teresa and Martin Luther King. Yet as her recently published diaries reveal, Mother Teresa was a woman driven by deep pain who believed that she was hated by Christ.[9] She also admitted to not having faith, a situation she knew was very ironic. It was these emotions that probably drove her to do her 'saintly' deeds - not out of the joy of serving, but to try and purge herself of the emotions that were dragging her down. Martin Luther King is also mentioned in saintly tones, but was repeatedly described by those who knew him as a very angry person.[10] Could this anger have been a driving force that led him to achieve such amazing deeds in the civil rights movement?

It all goes to show how there is really no judgement in our emotional dramas, in fact this emphasizes what masters we all are. We have created for ourselves exactly what we need in order to learn the maximum amount in our lifetimes and to drive us in our lives in exactly the right way. Debbie Ford, bestselling author of *Why Good People do Bad Things* writes in her

Amazon.com blog about how success is the best revenge - how she is well aware of the demons that haunt her, but how they have driven her to be successful and therefore do the service she was born to do.[11,12] Emotions are an integral part of our Genius Groove.

However, we do not have to be blindly driven by them. I acknowledge that part of my drive to write *Punk Science* was an 'I'll show them' attitude. I wanted to write the clearest book there was about science and spirituality to prove a point to all the people who have taunted me about my beliefs over the years. Interestingly, the more I worked on the book, the more the issue healed, until at the end, the sheer creativity and flow of it took over from the charged emotions and I was simply in a place of bliss and ecstasy - in my Genius Groove.

So even our emotional dramas play an important part in who we are, how our lives are shaped and why we are drawn towards our life's purpose, how we get into our Genius Groove. The amazing thing is - we choose it all for ourselves. As we shall go on to see in the next few chapters, we are more powerful and infinite then we can imagine.

An Exercise In Emotions

So far, we have discussed a lot of theory, but not done anything experiential. However when it comes to emotions, experience is everything; an intellectual understanding will not suffice. So I am going to discuss a tool that I have used myself occasionally that really helps to demonstrate these principles.

I discovered this technique when I turned up to work for what I thought was going to be a day of assisting in theatre. When I arrived, I was told I was actually going to be helping a particular consultant in clinic. This was unexpected and I had little time to prepare. This consultant was a tyrant who regularly hurled abuse and humiliation at his junior staff.

I was suddenly thrown into a state of fear, with all sorts of scenarios bubbling up in my mind. I had a few minutes before clinic so I just allowed all the projected disastrous scenarios to play out in my consciousness: the consultant hurling abuse, laughing at my incompetence in front of patients etc. After a few minutes of allowing myself to fully experience my fears, I no longer felt any. The clinic went smoothly and the consultant was completely charming.

I have used this technique on occasion and it is a fantastic one for when you are really nervous and haven't faced the reasons why. Remember that the more we push down on our shadow, the more it will bob up, so when we actually allow the energy of our emotions to go through us, a definite change happens to our beings. This is not an intellectual process - this is a vibrational change. To do this, we must fully feel our fear and emotions tangibly. By facing our fears completely, they can often dissipate and we are left wondering why we expended all that energy in keeping the beach ball under the water in the first place.

There are many books out there discussing emotional dynamics with suggested exercises. I am including a few in a suggested reading list at the end of this book.

Part Six - Time after Time

Chapter 12

The Time of My Life

"Well we know where were going

But we don't know where we've been" [m]

Talking Heads

When Tomorrow Comes

Now I am going to discuss one of the most controversial aspects of this book. In previous chapters, I discussed the Black Hole Principle and described how light comes through the multidimensions and into our dimension where it splits into a particle of matter and a particle of antimatter. I also stated that the world of antimatter is one of negative time where time runs backwards.

This means that all events in our world of matter have already happened in the negative time, antimatter region. We cannot access this information easily because it is hidden from us in a different level of consciousness.

This means that there is a dimension of the universe where all of our actions have already occurred! This is radical stuff, because it implies that our lives are predetermined and we don't have any choices. However, if you recall, the positron, electron and photon are in a continuous cycle with each other, so there is more to the picture. The antimatter and matter region are actually two parts of a whole, as the section on emotions described.

In a way, the matter and antimatter states are really light in disguise. In the big picture, the separation has not occurred and everything is still united. So the matter and antimatter states are also light at the same time. It depends on how you look at the situation and, as we view the situation from our 'matter world' perspective, we view the universe through 'matter-tinted glasses'.

In the big picture, light is never really separate and is the underlying truth behind everything in the matter world including ourselves. According to the Black Hole Principle every single part of us at every level from atoms, to chakras, to DNA, is creating from infinity. This means that a part of us exists way beyond the physical dimension that we can perceive with our senses. We exist in many dimensions all the way to the infinite oneness.

It is the infinite part of ourselves, out of space and time, that makes the choices. So although it seems that we have no choice from the timeline perspective, the infinite part of ourselves is always making choices from these infinite possibilities, placing these choices lovingly into our lives as a chance for our maximum growth.

Best of Both Worlds

This is how we can have the best of both worlds. Philosophers throughout the ages have been arguing whether we are ruled by an inevitable destiny or we have free will. The answer is both - we can have our cake and eat it. We no longer have to struggle to reconcile the existence of the timeless eternal state with the fact that we all grow up and grow old - we don't shrink back to being toddlers or even flip in between. No matter how much physicists and great sages alike try and convince you that time does not exist, the human experience is one of the passage of time.

The issue of time has also been hotly debated by physicists for the last century or so. It used to be an easy topic - it was assumed that there was a sort of universal clock ticking away in the background: that time was the same no matter where you were in the cosmos.

Einstein did away with all that when he showed that time behaved in a different way according to where you are in the universe thus doing away with the universal clock. So a twin who spent ten years in orbit in the Earth's atmosphere before returning, will have, in most scenarios of this 'twin paradox', aged more slowly than the twin back on Earth and the clocks on board the spaceship will be running behind the clocks on Earth.

The inference from Einstein's theories look much like those of the ancient sages: that time is an illusion and does not really exist. Yes, at a deeper level this is true and always has been. However, that does not help you catch the 6 o'clock bus back from Sainsbury's on a Thursday evening.

How do we reconcile the world we live in with these deep philosophies without getting totally stressed out? It is by realizing that we are 'both - and'. As described in previous chapters, the speed of light is not really the speed limit of our universe, it is simply a reflection of the vibrational limit of our consciousness of the three dimensional world. The world above the Perception Horizon is the world of massless light without charge. The world below the Perception Horizon - the world of matter is a world of mass and charge. The world of antimatter and matter are linked, although normally, we are not aware of antimatter - hence the big mystery amongst scientists as to where it is. Antimatter is not in the domain of massless chargeless light, but nonetheless is in a zone that is slightly too fast for us to detect with our normal senses. This is the one of antimatter or the c^2 region.

Everything that is in the matter world is in a relationship to the Perception Horizon that lies at the speed of light, which, now we have BHP, we know is not the ultimate speed limit of the universe. The mass of any object in the matter world is related to this speed. The matter world is the realm where time does exist and moves forwards. The universe is all about perspective. From the perspective of below the speed of light, time and mass exist. But as we have seen a few times in this book, this is not the whole picture and there is also the antimatter region. If we were to vibrate fast enough to match the c^2 region, we would find ourselves in a world where time runs backwards.

If we then moved our consciousness to beyond the Perception Horizon of both the c and c^2 regions into the timeless, massless state, from this perspective there is only The Eternal Now. In the big picture, time does not exist at all, the two timelines have cancelled each other out. This is actually the deeper reality and Einstein and the great sages are right. However, they have not reconciled this statement with our daily experience.

It is as if we view reality through the lens of whatever vibration you find yourself in. From the perspective of the matter world region, time seems to move forward; from the perspective of the antimatter world, time is moving the other way. Everything that we call our future already exists in this region.

In the big picture these two regions cancel out. It is all about from which vantage point of consciousness you find yourself in. So from our vantage point, time seems to move forward and this is the game we play when we are in the matter world. To deny the passage of time, as we are encouraged to do by some teachings, can practically give you a psychosis as you try and deny that you live in a world in which you grow older and eventually die. Yes it is true that time is ultimately an illusion, but if we accept the existence of a timeline in *this* dimension it will save a lot of tying ourselves up in knots trying to deny what we see around us.

Just An Illusion

A lot of the spiritual teachers who espouse the living in the present moment eg Eckhart Tolle, are providing profound and valuable teachings that the majority of people on the planet need right now.[1] But the *Power of Now* should be renamed the *Power of Presence and Conscious Awareness*, because it really is about being present to what is going on in each moment.

That time is ultimately an illusion is a very important realization for the majority of people on this planet who still haven't woken up to a deeper reality. However, to say time does not exist at any level of reality, just makes everyone stressed out. In an issue of *O magazine*, Eckart Tolle tries to answer the concerns of readers, who have commented on the commonly held experience that babies grow into adults and not the other way round.[2] Tolle dismisses these concerns by simply stating again that time is an illusion. This is not very helpful.

Although it is essential that people see through the illusion that they have been laboring under, it is also not very helpful that you deny people's everyday reality. To make peace with both time and timelessness occurring at different perspectives of the universe, paradoxically relieves some of the stress of the spiritual journey.

Although spiritual teachings are very useful, it is not useful to be in denial of the inevitable - growing old and death. Yes ultimately these issues are illusions in the big scheme of things, but not in *this* dimension and in *this* perception of consciousness. This is not to downplay the excellent teachings about being present and aware to what is going on in your life in each moment.

Hopefully, this clarification afforded to us through Black Hole Principle, will help stymy the flow of New Age bores who claim continuously that they are in the Now and do not want to be knocked out of it with other people's 'bad vibes'. There have been many an Eastern Guru who have claimed that time does not exist, but who ended up growing old and dying anyway. Our denial of the passage of time is really not helpful and causes more hassle than necessary. So thankfully, at last, we have a scientific theory that allows for this.

Chapter 13

Same as it ever was

"I'm the same boy

I used to be"[n]

Steve Winwood

Danger Zone

Now we get to the really difficult bit which is discussing how the future has already happened. Lots of people will not like this idea, as they feel it takes away their free will. And besides, doesn't quantum physics tell us that we choose our own reality from infinite possibilities? This is the section where you either sink or swim with the Black Hole Principle. But it is also key to a deeper understanding of The Genius Groove.

If the c^2 region really has a backward timeline then the future is already in existence and our fate is sealed. OK some of you are really running for the door! But if you look deeper, there is lots of evidence from the scientific world, the esoteric world and from your own life that seems to indicate that the future has already happened.

Is this even a question that can be scientifically verified? Well actually there are signs that quantum effects can have backward causality so that the future affects the past. Quantum entanglement happens not only through space, but also through time.[1] Loop quantum gravity, which is another leading physics theory, also speaks of the fact that there is a negative timeline.[2]

In fact, there are a few theories that allow for this and even some that link the negative timeline to the antimatter region.[3]

Save a Prayer till the Morning After

In terms of this backwards causality, where an event in the future seems to be determining the past, there have been some interesting studies done in clinical settings. For example one randomized controlled trial looked at outcomes in patients with a bloodstream infection in which a certain intervention was carried out on a test group and compared it with a control group.[4] Except in this study, the intervention was not a pharmaceutical agent, but prayer and, in another twist, the patients had been discharged years ago.

There have been a few studies which have examined distant intention or prayer in medicine, some showing that prayer has a significant effect. However, most look at outcomes that have yet to happen - this one actually had people pray for patients after they had left the hospital. Because it was a double blind study, the people praying for the patients had no idea as to the patient outcomes. Nor did the people conducting the study. However, the people who were prayed for showed a significant decrease in the time they had to remain in hospital. The weird thing was they had actually been discharged a few years *before* the prayer study was carried out.

American Science

In another type of scientific study, physicists have studied the principle of clairvoyance. In the 1970s, Russel Targ and Hal Puthoff were commissioned by the CIA to investigate a phenomenon known as Remote Viewing.[5] This is basically another name for clairvoyance which is the gift to see events and places without being in the vicinity or timeframe. People who are gifted in this ability can apparently view places they have never visited and give accurate descriptions of them. During the Cold War, this sparked the interest of intelligence agencies and the US government allocated funds to this project for many years: such was its importance.

The Remote Viewing studies done by Targ and Puthoff with their various collaborators, has left us with a solid body of work that raises some interesting points about the nature of time. Some of the results are still classified material with the US government, but Targ has discussed one scenario with the public which has implications about the nature of time.

In this scenario, one researcher remained in the laboratory with the remote viewer whilst another researcher travelled out in a car with a test volunteer. This person was instructed to drive the car, taking a series of random turns without any planned destination. The remote viewer back in the lab had to describe the scenes that the volunteer was witnessing.

On one of these occasions, the remote viewer started to described a scene involving a jetty and drew some boats. This was indeed the ultimate destination of the person driving the car. The twist is that he did this a whole hour before the subject had got there and, because they arrived at the destination by taking random turns in the car, the test subject themselves had no prior knowledge of where they were going to end up.

The study was done by physicists under scientific conditions which meant that the test subject and the remote viewer did not have a chance to confer before the study. This study implies that the future is able to be witnessed, as if the information is available somewhere if you know where to look. There are now various published scientific studies in existence that imply that information from the future is accessible.[6]

Picture This

There have been various mystics throughout history who claim to be able to see the future. One of the most notable is American Edgar Cayce (b.1877 d.1945) who was able to make astounding predictions and perform healing on people whilst in a deep trance state - hence he was known as the 'Sleeping Prophet'.[7]

These predictions, which numbered in their thousands, have been carefully recorded and catalogued so we have good records of what happened during these trances. Such was the reputation of Cayce that many important politicians and businessman came to visit him. He claimed to be able to look into the *akashic* records, a Sanskrit term for sky or ether. Via the theosophical movement, the term has come to represent a kind of cosmic library that contains all the information of the universe: a record of everything that has ever happened or will happen.

As we have seen in previous chapters the modern scientific concept of the Quantum Vacuum, through its holographic properties, also contains the information of the whole universe in every part, making it the modern equivalent of the akashic records. Modern philosophers and scientists see the Quantum Vacuum as a container of all the knowledge of the past, present and future of everything that has ever happened or will happen to everyone on any dimension. Theoretically, it should be possible to access this information as we are able to interact dynamically this field.

For some people, the talent of looking into the Quantum Vacuum and its holographic field of information is their Genius Groove - we call these people mediums, psychics, remote viewers and sometimes even prophets. However, we all have our signature vibration which is able to interact with the field and retrieve information from the universe that is unique to us.

Sometimes, the process of taking information from the Quantum Vacuum is called channeling, but really, because our brains are always pulling consciousness through from the universe, all thoughts may be referred to as channeling.

Our true moments of genius and inspiration are those that occur when we are so centered that we reach our higher aspects more clearly. These are the moments beyond space and time when we hear the voice of our eternal selves and go beyond our daily emotional lens. We become totally aligned to our true selves. This is the place that some people call the flow, or being in your element and what I am calling The Genius Groove.

For some people it is their Genius Groove to see deep into the Quantum Vacuum. These are the people who claim to see into the future. If the future is not already determined then how is it able to be viewed by either remote viewers under scientifically controlled conditions or by the many mystics throughout the ages?

Even famous prophets of the future such as Nostradamus and Edgar Cayce found it unpalatable to say that the future is determined and railed against it, advising people that their own prophecies were mere possibilities which could be changed by human choice.

Often mystics do not explain why their prophecies remain as possibilities when they seem to be witnessing the future so clearly. Indeed some of their prophecies, as in the case of Edgar

Cayce, can be corroborated later on. Edgar Cayce predicted the Depression of the 1930s for example.[8]

What determined that Cayce sometimes saw a possible future and sometimes the 'real' one? There is no satisfactory explanation and it also makes no sense. If mystics throughout the ages are able to witness the future by accessing the information of the Quantum Vacuum as an aspect of their personal Genius Groove, then this implies that the future has already happened - that it is determined.

Big Decision

At times, we all get a sense of fate and destiny playing out in our lives. Many people have had the experience of being thwarted in some endeavor only to find out that if they had been successful, they would not have followed what was ultimately the better path. This is when we glimpse a greater plan that has already been laid out for us. After a while, our focus shifts and we stop pushing from the past to make something happen, but tune into our future to see what has already happened and take action in the present moment to allow that future to unfold.

I call this 'living from $c^{2'}$ after the region of the universe that is running backwards in time according the Black Hole Principle. When we live this way, decisions we make are different: created not from logic, but from using our intuition to feel what we have *already* done.

Living in this way is also is very liberating, because you know that you can never make a wrong choice; the choice was already made. So whatever is right for your life is *always* unfolding. It doesn't mean that you don't have challenges, but you understand that you have created this challenge for your-self as the ultimate loving gift to yourself and that is liberating and enlightening.

When you are 'living from $c^{2'}$ you can recognize someone from your future. I mentioned a good example in *Punk Science* when author, Joseph Jaworski met his future wife in a crowded airport and knew instantly that this women was of importance to him.[9] She too had a feeling that she was going to meet some-one significant. They instantly recognized each other - not from the past because they had never met - but from their future.

Is It A Dream?

In my own life, this principle has been playing out beautifully. My family in particular have watched as events unfold in my life revealing a divine order and a sense of the pre-ordained. Without me insisting on marrying my ex-husband and not just dating him, we would have split up a few years before we did. He was more influential in my life as my husband rather than a boyfriend in terms of demanding mental health checks etc as legally he was my next of kin. This led to the GMC investiga-tion which then liberated me from the medical profession in a way that gave me the time and space to write *Punk Science*.

If I had tried to tell my family that everything was working out according to a divine order, they would not have believed me, but because I actually lived out these events, others can witness this sense of order and see that everything works out exactly how it is meant to.

Another incident in my own life which demonstrates this sense of destiny is the way in which I met my current partner, James. Months, even years before we met in person, I saw him in my dreams and sometimes in visions, which I recorded, so I can actually look back on them now. Even though I sensed someone new was going to come into my life I was adamant that I wanted to find a publisher for *Punk Science* before I met him.

Sure enough, a few days after I was signed by a publisher, I met James. I had actually spoken to him a few times on the phone and I knew very early on that we were going to be together, even though I didn't even know what he looked like. I actually told my friends that I had met someone and would be moving up to Derbyshire. I even arranged my speaking schedule around this - such was my sense of my future destiny.

A lot of us have a sense of destiny and fate about certain incidents in our lives, but many esoteric teachers are caught up in the paradigm of total free will at all levels of reality so either dismiss this or give the ludicrous explanation that the big things in life are determined, but the little things are not. So what determines what is a little event and what is a big event?

For someone like myself who has seen the multidimensional perspective many times, there are no small events. What we perceive as little movements such as picking up a fork to eat lunch are translated geometrically through many dimensions and have huge significance. Everything you do is important and is part of the intricate fabric of your life that meshes all the other dimensions aspects of you. There is no small stuff!

You are an infinite being who has aspects throughout the cosmos, do you think that eating your lunch would not be significant? When we start seeing our actions in this way, it gives us a whole new perspective. In an age of celebrity obsession, this perspective could not be more timely. We are all megastars and if we were aware of the higher dimensional perspective, we would know it!

There are no insignificant events in our lives which means that all aspects of our lives are determined and meant to be, not just what we perceive as the 'big stuff'. There are going to be many people who will be very unhappy with this, because it appears to take away the locus of control. Some people also will feel more determined by this and want to change their life and make something of themselves. They will feel angry by these words and rail against them.

This reaction too, has already been determined as these people will do something unusual in their lives and are sensing their own futures. Please note that unusual does not mean better, it may be your genius to bring up a child which may not need you to create unusual circumstances. But if it is your genius to be a cabaret dancer, it may take some different circumstances to help you realize this, as it is not such a common career.

Don't Give Up

As a sixth form student, I travelled to London to attend a conference at The Royal College of Surgeons designed to help school kids learn more about medical life. As part of the day we met up with a 'real life' doctor. My friend and I sat and listened to a jaded, tired, overworked registrar who spent half an hour telling us not to go into a medical career and about the terrible conditions of the job. (I later found out that he was right!)

After this barrage, my friend turned to me and said that she was no longer going to apply to medical school. I was stunned. The registrar's vitriol against his career only served to strengthen my reserve and make me more determined. True to her word, my friend did not pursue a medical career and, when I got a place at a London medical school, I had an overwhelming feeling of it having been meant to be.

If I say to people that their lives are determined, some may interpret this as saying - put up with the drudgery there is no point in trying to change it. But just like my determined self who became even more set on a medical career after someone tried to put me off, some people have actually already decided to change themselves and lead an unusual life. Their changing themselves has already been determined too! And of course the choice is made by our infinite selves.

People who decide to change sometimes visit healers, life coaches, teachers or even read a book like this, to help them transform. If you enter into this sort of dynamic it is important to remember that an agreement has already been made at a soul level for that teacher to come into your life to make the transition necessary for this moment. It has already been determined and the different choices that you make as a result of that transformation have also already been determined.

You put those systems in place to make those changes in yourself that will lead to a different type of life than you were having before. You were always going to meet that teacher and make that change. You are therefore a master of your own life and so is your teacher/coach/therapist a master of theirs.

Too Much Too Young

A few years ago there was a show on the VH1 channel called *The Rise and Rise of...* and each program examined the early history of pop stars before they became famous.[10] The majority of the time was focused on their rise to fame and not what happened afterwards.

One of the programs featured the singer, Britney Spears who rose to fame at the age of fifteen and has been one of the world's most famous celebrities ever since. It was fascinating to see her early life featured on the show and what she was like as a child. Bear in mind that by the time she became famous, although she was young, she was already an accomplished performer who must have practiced a lot in childhood.

In film footage and stories from her childhood it is clear that the young Spears was putting in an extraordinary amount of effort into her performances - this child was driven. There were stories of the young girl asking her peers to rehearse a dance again and again, even when they were not interested. She demonstrated an amazing force in such a young girl; did a part of her know the future and was preparing for it? A lot of people would argue that she practiced so hard that eventually she made it. She pushed for it to make it happen.

But why weren't the other girls rehearsing so hard? Because they sensed their futures were not as pop stars but hedge fund mangers, doctors and mothers. People practice hard, partly because they have found their Genius Groove and partly because they sense their future. The part of Spears that is out of space and time was sensing the future that she would be a star and so practiced hard and therefore had the polished talent to get her noticed. It is this circular causality with the future influencing the past that made her a star!

The King of Wishful Thinking

Don't people just have lucky breaks or the right circumstances? Well, guess what! We are the ones who have placed these circumstances into our own lives so that we can follow the futures that we have mapped out for ourselves. Malcolm Gladwell in his book, *Outliers* puts a case forward that genius is not born, but a result of certain circumstances.[11]

He breaks down the stories of the heroes who, on the surface look as though they have broken through all odds and argues instead that they were actually at the receiving end of some happy coincidences and occurrences that enabled them to be successful. He includes everything from birthdate to get into the best hockey team, to living in the right area at the right time e.g. Silicon Valley in the 1970s for Steve Jobs and co, to having the right person notice you at school. He is trying to put forward a case that there is nothing particularly special about successful people - they just got lucky!

However when you start to examine some of these stories, these people tend to 'get lucky' over and over again. The story of Bill Gates' success reads like this.

• He was born to wealthy parents who could afford to send him to an elite private school.

• This school was bought a cutting-edge computer in 1968 by the forward thinking, fund-raising mothers of the pupils and the young Gates joined a computer club.

• One of the pupils at the school had a mother who was a programmer at a local computer firm and asked Gates to test company software at weekends in exchange for programming time.

- When that company went bust, Gates happened to know of another local company with a similar arrangement based at the University of Washington.

- Gates happened to live within walking distance of the University of Washington.

- The university happened to have free computer time in some departments in the early hours of the morning which Gates and friends found out about and took advantage of.

- They happened to know someone who was asked by a company if they knew of any programmers.

- It happened to be that the best programmers that this person knew for the job were the kids who were clocking up hours on the university computers - Bill Gates and Paul Allen.

- The school happened to be the type which would allow Gates to miss a term working for a company and getting exactly the sort of experience he needed to form the basis of all that he did next.

That's quite a list of coincidences. It starts to look suspiciously like a plan! But in the case of Bill Gates and many others, the plan would have to go beyond childhood to before birth as his mother and father are so pivotal to the story of his success.

If the future is already determined by our higher selves - this would fit Gates' story and his suspiciously long list of lucky breaks. He chose the right parents. As do all of us, we choose all events for our lives and place them into our future timelines in order to learn what we need to learn in our own lives.

Malcolm Gladwell also argues that if you have to put in 10,000 hours of practice at any talent or skill before you excel at it. But he doesn't explain why some people put in those hours and others don't, are they just lazy? Or is it the Britney Spears effect again? Some people put the hours in because they have found their Genius Groove and they can feel that it is a part of their future even if it is totally subconscious.

If we take Gladwell's example of playing the violin, why do people practice more than others? Gladwell deftly puts all violin players into categories of accomplishment according to the hours that they practice, but doesn't explain their motivation for doing so. Could some people be practicing more because they are in their Genius Groove?

Gladwell even seems to dismiss Mozart's talent as just being down to having twenty years' experience of music. In that case then everyone with twenty years' experience would be at as high a level as Mozart, but it seems his gift has not been repeated. Was there something about Mozart's relationship to the Quantum Vacuum that reflected him and only him?

And the success of The Beatles, Gladwell attributes to the many hours that they played in Hamburg which honed their skills. But which out of the many bands playing in Hamburg at the time can we name now?

Gladwell has hit upon a really important point, that nobody makes it to the top without the right set of circumstances that allows that. But what he doesn't realize is that those events, that he dismisses as a series of coincidences, have actually been planned by the infinite self from beyond space and time in order to create the exact circumstances to learn exactly what is needed in this lifetime.

So if the future has already happened, why don't we all just go home now? Why do we bother with life if everything is already determined? Let's say your future *is* already determined - do you know what it is? No! Not even the most active clairvoyant knows everything there is to know about their own lives. And this is why life is still a learning process - although we may know the contents of our lives at a higher level from our consciousness beyond the speed of light, at the level of our being below the speed of light, we can call this the ego self, we do not know the contents of our lives in full. So our lives remain a fascinating journey with many learning processes that remain with our souls even after death. It is not money or wealth or possessions that we take with us beyond death, but how much we have expanded our consciousness by all the lessons we have learnt.

That is our true wealth and our true currency. When we have left our bodily form, we no longer have a need for DKNY! If you understand that your consciousness goes beyond this dimension you can see things in a different perspective. So even though we know that our future is determined it doesn't mean that we know what it is. By understanding the bigger picture however, you realize that everything large or small in your life has been created by the aspect of you that is your 'Higher Self'. You are not a victim in your life. Every choice you make is the right choice and every incident you have set up is to give yourself the most loving lessons in life through learning exactly what is right for you.

I really feel that this knowledge helped me to survive what I have been through. At the very moment that a psychiatrist and

GP came to visit me with the intention to see if I needed to be taken to a psychiatric ward under the Mental Health Act, I knew that I had tried everything to stop this situation from happening. Yet still it happened.

I knew then to relax because I realized that this was my own doing, my Higher Self had set my life up like this. Even as I was being referred to the GMC and was experiencing a lot of pain, I also knew in the depths of my soul that this would be the making of me and even said as much at the time. This did not mean that I did not experience heartache, but the knowledge that the future has already happened and that we create everything in our lives from a higher level got me through those troubled times.

I can tell you that a lot of people without this knowledge would not have made it through such an ordeal. Many doctors who are facing any sort of GMC investigations, even when they know they are innocent, have actually killed themselves rather than face such trial and humiliation. If it were not for this knowledge, which I had of course decided to receive before this event, I don't think I would have made it.

So I can speak from my own experience that this knowledge can help you in the most dire of situations that otherwise make no sense. I now live in a completely different way to before. Instead of thinking that little things in my life are random and that I am a victim of circumstance, I know that I am setting myself up to learn something. The learning is the treasure. Decisions become less nerve wracking and of course you have absolutely no regrets. It opens you up to the beauty of a new way of being, something that I call New Paradigm Living which I will discuss at greater length later. First let's have a look at what the

Black Hole Principle means for something called the Law of Attraction.

Chapter 14
Please, Please, Please let me get what I want

"Shoot it in the right direction

Making it your intention

Live those dreams

Scheme those schemes" [o]

Frankie Goes to Hollywood

<u>The Key The Secret</u>

There is a big interest right now, after the release of the best-selling book and DVD called *The Secret*, for something called the Law of Attraction.[1,2] This is the idea that your thoughts and intentions can actually attract things into your life. This is not really a new concept, but *The Secret* came at the right time for many to be ready to receive it as well as being accessible and cleverly marketed.

This modern incarnation of the Law of Attraction utilizes principles in quantum physics to explain how it works. I discussed the science behind the Law of Attraction and distant intention in *Punk Science,* so will not go into detail here. Instead I am going to discuss the way in which these principles seem to operate on a personal level and how we can gain new insight into them with the Black Hole Principle.

<u>Do Fries go with that Shake?</u>

Many people are now using this Law of Attraction in their own lives, hoping that they can get richer, get the Ferrari and the girl/boy of their dreams. *The Secret* has been out for a few years now - enough time for many people to find out a fundamental truth about the Law of Attraction - it doesn't always work.

People are frustrated that they are not manifesting what they want and wondering why it doesn't always seem to work. Actually, it *is* always working. You always get a life that reflects your consciousness - *all* of your consciousness and *all* of your being, not just your ego, which is the implication of some teachings.

Believe me, I used to be such a big advocate of the Law of Attraction. I had the vision boards, I even walked around with some beads and would chant, "I have a house in Berkhamsted, I have a house in Berkhamsted". As any Law of Attraction student knows, you don't say you want something when you are trying to manifest it, you say you already have it and walk around acting as if it is already here. The idea is that the universe will just form itself according to your resonance and deliver up what you are trying to manifest.

I did indeed manifest a few things that I wanted and, like many people, I noticed some things about the whole mechanics of it all. The Law of Attraction works best when you have no attachment to the outcome; when I have a casual thought something will manifest almost immediately.

One day I was getting ready to have lunch with my mother and I had a 'throwaway thought' that I needed to have a certain type of handbag to match the outfit that I was wearing. Half an hour later, I met my mother for lunch who brought out a gift for me that she had bought on her recent holiday - it was the exact handbag that I had momentarily pictured in my mind and then forgotten about. I had no attachment to the handbag so it came to me almost immediately. I have countless examples of these incidents from my daily life.

Unfortunately with most of our wants, we have a massive attachment to them because they are coming from our ego selves. Do we really need a Ferrari/ Gucci handbag/ bigger house? These tend to be ego needs and come with a sense of attachment. As such, these types of manifestations do not always happen and people get frustrated thinking that the Law of Attraction is not working. I also noticed, like many people that after some success initially, the Law of Attraction did not work in quite the same way in my life. What is going on?

I Got a Catholic Guilt

We are now in the post *The Secret* era when literally millions of people around the world are finding the same conclusions that I have and some are angry and feel they have been conned.
This has led to a plethora of advice books telling you about the 'Missing Secret' or the 'Real Secret' or 'The Secret they do not want you to Know'.

All of them are trying to tell you that something is missing in *The Secret*. Usually this consists of informing you that -
• You have to allow what you want into your life and not block the process.
• You should not think negative thoughts about what you are trying to manifest

In other words, 'You are just not trying hard enough, you bad manifester, or you would have that Ferrari by now, what is wrong with you?' It feels like the so-called New Age is bringing us right round to the era of Catholic guilt. This leads people to feel even worse about themselves for not being 'spiritual enough' to get what they want and that there must be something wrong with them, because their attempts to manifest are just not working. When this happens there is usually some smug New Ager around, telling them they are not trying hard enough.

Back in 2004, my friend Kim and I were organizing the first Children of the New World conference for that September in a hotel in Bedford, UK. A few months before that event, Kim had organized another event at the same hotel, on the subject of angels and gone to great lengths to publicize it. She worked very hard and was frustrated that very few people booked for the day.

I was dismayed to find her very upset in the approach to the angel day due to someone telling her she just wasn't asking her angels hard enough! So instead of feeling uplifted, she felt really bad. The day of the angel seminar came and there was a very poor turnout.

Furthermore, the hotel had not fulfilled their promises and provided enough staff. So lunch time came and the kitchen was overrun with orders that took hours to fill with only a young lad acting as waiter. He had been up all night already and was covering someone else's sick leave. The weird thing was, the seminar room was filled with angelic presences. An angel even moved across the room as I was doing a talk and managed to leave me quite speechless. We couldn't understand why, if all the angels were out in force, that there was such a poor turnout of people.

In the aftermath of the event, Kim was furious with the hotel staff. She leveraged the fact that they had let her down to get the most out of them for the children's conference being held later that year. The new manager showed up for his first day of work to find Kim on the rampage (in her gentle way of course!). She asked for more staff and attention for the children's conference and boy did we get it. Despite the conference being held on a Sunday, the staff were in very early and were diligent and attentive.

There were no problems serving lunches and the event itself was a resounding success. We could now see the bigger picture; the angels were helping make the second conference a success because it was more important at that time to be highlighting spiritual children, something that nobody had really done before in that way. The conferences ended up being such a success that the next year's events were filmed by UK's Channel Four and they have continued to generate interest ever since.[3]

Unfortunately, the sorts of comments Kim's friend made, that she wasn't manifesting hard enough, are all too common. It can be seen as a sort of 'manifestation snobbery' and is not very helpful. But why is it that sometimes manifestation happens the way we expect it, and sometimes it doesn't and, even worse, why do 'crappy' things still happen?

What if God was One of Us?

There is something deeper happening than the simple satisfaction of our ego wants - a higher force is in action. Most people don't believe in a sense of God as a white, long-bearded man in the sky, but believe instead in a higher power that unites everything. In fact, 'The Force' from the *Star Wars* films is closer to people's modern ideas of God.[4]

In our post-religious era, a belief that is rising in prominence is that everything in the universe is God coming from a unified source and that it is only an illusion that we are separate. People are fine with this philosophy until it comes to applying it to their own lives and themselves.

Suddenly there is separation. People think, 'But I can't be God. If I were God, why is my life so crap?' This is one of the reasons why the Law of Attraction has been so popular, because so many people cannot accept the world they are in and want to change it.

But even armed with the knowledge of the Law of Attraction, things still happen that people are not prepared for. And then they get frustrated and wonder what is going on. The idea that you have to manifest things in your life actually involves an ideological assumption - that your life isn't already perfect.

But if everything is God, then everything must be perfect including yourself and your life. Everything in your life is God: every single situation, even the ones you are not happy with. If you are not happy with your life as it is, it means you are not accepting you or your situation as God. As my friend, Story Waters says, *You are God, Get over it!* [5] The fact is, you have already manifested everything in your life already. You yourself have placed everything in your life even the 'crappy' things, in order to give yourself the maximum soul learning possible.

It is your non-acceptance of this fact that is bringing you pain. However, with our knowledge of the Black Hole Principle we can understand that everything is both infinite and in space and time at the same time. The part of you that is out of space and time is choosing your life: every little detail, for the maximum learning of your soul as it goes along its journey. This is the treasure you take with you from this dimension, not money or material possessions.

Higher Love

Once you understand that a higher aspect of you is in total charge, you can surrender to this higher power: *you*. The journey of manifestation and the Law of Attraction can go something like this.

Stage One. Initially it will work very well. This teaches a truth about the universe - that consciousness is fundamental to reality.

Stage Two. You will start to realize that the Law of Attraction does not work all the time, as so many of the fans of *The Secret* have already discovered.

Stage Three. You start to realize that 'crappy' things still happen and wonder, 'who ordered that?'

The process of discovering manifestation is huge for the planet and all of those teachings are very welcome because they help people to realize that consciousness is the ultimate, fundamental reality. To learn that your thoughts can attract material things is a major shift in awareness that is taking this planet into a new era. Once you make this connection you don't need it anymore as you go into the next stage.

Stage Four - The Stage of Surrender

You then can shift into the deeper reality that all is perfect and there is nothing to manifest - it is already done! This is the process of surrender. Surrender has some dreadful connotations in our society of giving in and giving up to someone else. It is associated with defeat.

In a way it is a defeat, but only of the ego to stop struggling against its reality and to realize that the Higher Self is in charge and let go. When this surrender happens, the most extraordinary things can come into your life. Ironically all the things that you were trying to manifest will probably arrive into your life effortlessly and in a greater form that you could have imagined.

Heaven Knows I'm Miserable Now

Why do you manifest 'crappy' things in your life? Well you are not only manifesting from your conscious thoughts, you are also manifesting from your subconscious emotions. In fact, I believe that you are also manifesting from your antimatter region. This contains all the emotions and issues that are hidden from you. Life will bring them to your attention when the time is right so that you can heal them. They will come into your life in a way that will disturb you so that you take heed.

Actually this is the Law of Attraction in process, because it is the vibrations of your subconscious emotions that are attracting situations into your life. We see them as 'crappy' because facing your shadow is not always comfortable. But they are not 'crappy' at all - that is a judgement. They are actually great opportunities that we have placed lovingly into our own lives from the perspective of our infinite selves so that we can learn.

When something comes into your life that you are not so comfortable with, instead of ignoring the situation to focus on the 'good' things in your life, which is what your ego wants, you can realize 'hey, this is coming up because I have reached the next stage of my life and my own growth. What is there here for me to resolve?' In fact the lesson from your subconscious has probably bubbled up *because* you said 'yes' to the universe and committed to your own growth, but in order to get to the next stage of your life you need to heal something you are not aware of.

So something will come into your life that you are bothered by, such as a disturbing colleague at work, but remember this is always a mirror. When you realize what you have to learn in the situation, at an emotional level not just an intellectual one, then you move the situation into light as we have discussed before. Also it is not just situations that have developed from issues in childhood of *this* life that can be triggered. They could be from past lives or even events on different dimensions that have gone into your antimatter region and need to be resolved.

For example, my biggest fear was being investigated by the GMC. In order to move onto the next stage of my life and be free to say what I like without being in fear, I needed to deal with my subconscious emotions, which may have originated in past lives. Interestingly, the week that I went through the interview with the psychiatrist who was determined at the outset to diagnose me with a bipolar disorder, a strange mark appeared on my neck. It made me think of a noose, as if one had been around my neck and left a dark circle on my skin.

I kept this insight to myself, but as I was recounting the difficult interview to a couple of friends, one of them said, "what have you done to create this?" This caused me to shift gears and I showed them the mark on my neck. They instantly recognized this as a noose mark and made the connection that these were past life issues that I was dealing with.

What exactly these issues are, I still don't know, but I have come across many people who have multidimensional awareness who also have to deal with past life memories that involve being persecuted for their beliefs. These past life memories are often buried, but nevertheless, people are held back by them and are afraid to speak up in their present life about their multidimensional, spiritual nature in case the same thing happens again.

Hooked On A Feeling

It is not only the actual issue that disturbs us - remember we are dealing with consciousness here. It is also the judgment of self that attracts something. So even if you do not behave in the way that you see in the mirror situation in front of you, you may be worried that you do. This actually creates that quality in front of you so you can see it fully, assess your own judgments and let them go.

During our lives, emotions from our subconscious selves will continually bubble up to be healed. They sometimes manifest as difficult situations in our lives. If we face them and heal them, we can transform the situation into light, but if we don't, we will keep attracting the same issue and continue to feel weighed down by them.

So in fact, if 'crappy' things are happening - give yourself a pat on the back, you are making progress; you are looking more deeply into your own soul. And if you are wondering 'who ordered that?' Realize that *you* did in your infinite love and wisdom for yourself. Even though whilst we are on the Earth plane we sometimes want material possessions, our real riches are the lessons of the soul. We will be able to take the accumulation of knowledge with us to the next dimension, because consciousness is our fundamental nature.

It is the quality of your life experiences that counts when you move on to the next level of existence. This fact was discovered in amazing circumstances by a man interviewed by Oprah Winfrey in 2007.[6] He had been in a 'plane crash and had witnessed the deaths of many people. He saw what he described as an 'aura' leaving each body as each person died and noticed that some were brighter than others. He knew that this was due to the quality of each person's life and after his ordeal, he resolved that he would have one of the brighter auras by improving the quality of his life.

So it is *you* who is manifesting the seemingly 'crappy' things in your life because in your infinite wisdom as God you are planning your life for your own maximum growth. If you truly understand this concept, you can stop denying the things you do not like in your life and and start transmuting them. You can start understanding why they have come into your life and accept them.

To find out why you have brought something into your life, you need to look inside yourself for the vibration: the memory that is matching the situation. Once you have found what the situation is teaching you about yourself, you can then shift your awareness to see that the situation was healed all along and move it out of polarity and into light. This stage that can be tricky, as it means actually discovering the emotional vibration within yourself. That is where it sometimes helps to have the assistance of some type of therapist, although it is not essential.

There are many different healing modalities out there from NLP, to The Journey™, to spiritual healing that deal with emotions. There are some, such as Emotional Freedom Technique (EFT), that give you a way to circumvent actually feeling or reliving your emotions. These therapies are all good, but if you don't understand or experience the processing of emotions in their fullness, these shortcuts will not work all the time. Understand the long hand before you understand the short hand or sooner or later, an emotion may bob up for attention. If you use these techniques and say, "Hey, look at me! I'm healed and I don't have to worry about understanding why I was in pain the first place - clever me", nature will show you after a while that the ego won't be able to get away with it that easily; we are here to learn.

It is important not to beat yourself up about having emotional 'issues' and having something to heal. The idea of life is not that we come into existence as fully healed beings. We have lessons to learn, but we don't have to resolve our emotions all at once. Our higher selves have determined when it is that we need to resolve something - everything happens at exactly the right time. So if you don't feel ready to face something, take it easy. If you are ready to shift, life will tell you, by putting the issue in front of you until you need to face it. But don't beat yourself up or think you are not growing fast enough.

It's Got To Be Perfect

This concept of perfect timing is so profound that when you really take it on board it has the ability to transform your daily existence. I no longer get irritated at traffic jams, but know there is a reason why I am stuck behind a lorry. I trust that the time I arrive is always the right time - for whatever reason. Sometimes it is to meet someone, or miss an event so I have to go to another. Sometimes it is to feel what it is like to be late as there is an emotional lesson in that. I have shifted into a way of life where I not only know intellectually that everything is unfolding perfectly - I embody this principle.

In terms of cosmic ordering - yes you are ordering. You have already ordered. When you live a life of surrender to yourself you start to flow with the tao. This absolutely does not mean that you do nothing and just laze around. It means going with the impulses of what is within you and seeing pointers on your journey in your environment as well. Something will tell you the way, and if it isn't clear immediately, wait and life will show you.

Since 2002 I have lived a life of surrender. I often say, 'is that what life wants me to do now?' because I have given so much control over to my Higher Self. After all, my Higher Self had the wisdom to create an amazing orchestration of events in order for me to make a profound, new scientific discovery. However, if you were to ask my ego self at the time when I was losing my job, it wasn't too happy! I just had to trust and surrender that all was perfect even though I didn't know why at the time.

I also noticed that what really moves you ahead on this journey is something I call 'polishing the crystal'. It involves doing the inner work, really facing your shadow self. After a while, it becomes so second nature, you are actually doing it instantly in the moment. Eventually you do nothing at all as you understand at a deeper level that you have already manifested everything and that all is a reflection of your vibration at some level. Your life starts to then look quite ordinary as there is no big song and dance about being spiritual. Your life becomes fully integrated on all dimensions, 'after enlightenment: the washing up', as they say.

By resolving your emotions, you move your life forward. You actually change your vibrational frequency and this attracts a different life. So if you are looking to attract more of the life you feel you must have (because you remember it from your future) face your shadow and do so with an unattached and honest heart and you will find that you change your frequency. This will attract more of your deeper self towards you. This is the true Law of Attraction.

Having explored our emotional lives, we will now explore the next steps of getting into The Genius Groove.

Part Seven - Into the Groove

Chapter 15

I Want to Break Free

"I traded fame for love

Without a second thought

It all became a silly a game

Some things cannot be bought" [p]

Madonna

True Colors

Let's just recap for a moment on the subject of genius. We have seen that everyone has the potential to be a genius and not just a few individuals. We have discussed how the changing science of consciousness, from biology to physics to neuroscience, has given us a new understanding of genius. We have also looked at how the financial and school systems are set up to keep us away from our genius and how emotions are the key to getting deeper into your authentic self.

So how does one actually get into The Genius Groove? Most of us actually have no idea who we are, let alone what our true genius and passion is. Usually, by the time someone is an adult in Western societies they have lost touch with their inner spark. All the concepts that we have about the world as a child have mostly been quashed and we no longer remember who we are and what we are about.

It is hard for us to even begin to recover our authentic selves and most of us are actually too afraid to step off the treadmill of our lives because we fear what will rise up from our subconscious. You see although there are forces orchestrating our finances so that we stay in struggle, some of us are actually happy with the status quo. For many, truly stepping into an authentic place of creativity is very frightening.

The Reflex

I have not been immune to these fears either. For years I hid my true passions from the world and even from my husband at the time. When the truth came out, it was such a tremendous shock and explosion that it had quite dramatic consequences. Once things settled down, I realized I had previously been out of integrity with myself.

Despite the difficulties along my journey, I am much better off for having it all out in the open. Furthermore, my career has reached new heights that would never have happened had I stayed where I was. So I speak from the first person when I say that it is a scary experience to step into your true genius and your true power. It can bring up many hidden fears in both yourself and others around you.

We feel it is easier to stay where we are rather than face the deep dark recesses of our own minds and our emotions. We tell ourselves that it is only 'other' people who are exceptional and can achieve more than a mundane life. We make every excuse to stay safe. But it is not safe to stay put. In fact it can be pretty darn lethal.

I worked for some time at the Bristol Cancer Help Center (now called the Penny Brohn Center). This is a wonderful place, funded by charitable donations, which helps people with cancer: using empowering meditations, complementary therapies and counseling. For a while, I was a holistic doctor there and worked on the two-day courses designed for people with cancer and their carers.

The experience was an immediate eye opener. Just the atmosphere alone was healing. For people who have been living with cancer, being somewhere where you can actually talk about cancer without creating awkward emotional reactions is a real joy.

Part of the course involved a lecture on mind-body medicine - the science of psychoneuroimmunology. This is a relatively new science that has emerged from centers such as Rochester University in the USA. Research in this new science shows that people who have the type of personalities where they are always putting other people first are most likely to develop cancer.[1] Time after time, I saw people's jaws drop as they suddenly realized that the sort of behavior, that they believed reflected being a 'good person', was actually making them sick and in some cases was going to kill them.

A lot of them had put aside their own dreams to look after their family or suppressed their own creativity in order to do what is expected of them. The trouble is, the soul fights back so although we can get away with this for a while, over time these soul impulses to be your true self rise up and fight back, sometimes through the body, giving a message to change course.

There are now many books about mind-body medicine by excellent authors such as Dr Larry Dossey who uses these sorts of principles in his own medical practice.[2] A few doctors are now trying to find the connection between subconscious emotions and physical symptoms in order to heal illnesses and there have been numerous scientific studies to show that this approach works.

So some illnesses can be seen as a message from the body that the person is not being totally authentic in life. Sometimes the symptoms may not be physical. In my own medical practice, I found that many people with depression or even minor sleep disturbances were actually unhappy with their career and knew they needed to be doing something else.

I learnt pretty quickly by observing my patients that there is no escape from your true authentic self. If you deny who you are, there are consequences, which may result in physical or mental illness.

Back to Life, Back to Reality

In our society we are not taught to understand these signals. Our society values the physical dimension above all else. Even recurrent dreams containing messages from the soul are seen as sidelines, only relevant for 'soft' sciences such as psychology, but even in these disciplines there is a shift towards categorizing all mental phenomena in a way that can be treated by pharmaceutical agents.

This is all part of the approach of our society to deny there is anything but the physical. We are encouraged to stop listening to the messages that come from within, to deny the soul. The same message is reinforced countless times in our lives. Just look at the next commercial break on TV to see how many times the idea is reinforced that fulfillment comes from buying a physical object. By the time a child has left school they are usually out of touch with their own inner promptings. Factors like the media, commercials, the established systems like education and medicine and their own circle of family and friends encourage this shift in consciousness away from the center.

Hungry Like The Wolf

It is our inner promptings that take us back to ourselves. Something can come into our lives that taps into our deep yearnings to be whole again and remember who we are. This is what I describe as The Call and it can come in many forms. It is a prompt to look at what we have forgotten, to move beyond the intellectual self and get in touch with our deeper emotional and authentic selves.

Of course, as I have explained before, we choose all of our lessons for our lives so if we never listen to The Call, that is OK too. Some people have chosen to experience a life without living their deepest urges and to experience the consequences of this. It is just that now we are in an era where this is happening less and that more people are truly wanting to become more authentic. In fact, you could say that collectively as a planet, we have all been on a journey of losing our authenticity so that we can truly know how to find it again.

There are many ancient cultures that speak of a Golden Age long ago and that we have moved away from that era into darkness, but we are headed back again as the Earth moves in cycles.[3] People are looking to the year 2012 as being a pivot point in that journey back to a new Golden Age. As we approach this new era, people are moving into a new sense of connection, spirituality and authenticity, indeed it may be the very method by which we get there. So our personal journey is actually a part of a great cycle that is happening on this planet, from union, to separation and back to union again. There is no judgement in it.

There are many methods of how to come back to our true selves and it is a journey that never ends. So now let's look at some of the steps along our journey to The Genius Groove.

Chapter 16

Groove is in the Heart

"No one can be myself like I can

For this job, I'm the best man

And though this may be true

You are the one and only you" [q]

Chesney Hawkes

So putting everything that we have discussed into action, we now have a system of how to get into your Genius Groove. This is really a process of understanding who you really are. By doing this you develop a new relationship with yourself and your universe, one that is authentic: when you are flowing with the true reality of your heart.

At the time of writing these words, the world is going through tremendous change. Headline after headline describe how we are in the biggest financial meltdown that the world has ever seen. It no longer makes any sense to cling to the old you. You are in transition if you like it or not. The old world and identity no longer exists. Either you can try and cling to the old, or you can take this opportunity to find out who you really are - who is the authentic you.

This is a chance to dig deeper than you ever did before and discover your passion. The old rules that we previously slavishly followed, of getting a good job and keeping your head down, are disintegrating. There is no better time than to embrace the changes and sink into your authentic self. Here we go...

Promised Land

We have already discussed how the current education systems and other structures in the Western world suppress the true voice of who you are and put you into neat categories of employment. We have also discussed how the system is set up so that you will be a worker all your life, so that you will never be free and how very few people earn enough in the UK and to even own a decent home. Despite being in a country of apparent luxury, we are basically slaves in an invisible cage.

Because we don't know how the system is set up, we go through our whole lives thinking that if we work just a little harder, just conform that much more to get that promotion we will enter into a mythical promised land and be one of the 'haves' and not the 'have nots'. We have bought into so many *shoulds* and *musts* and concepts of what it is to be a 'good person' that we often don't question their validity any more.

But the human spirit is strong and after a while, up comes the little voice, the one we want to keep quiet and subdued by being busy. It says, 'Is that all there is? Is that all there is to being human and who I am? I'm not happy, I'm not living my passion.'

We try so hard to bury our inner voice and become a 'good' person as defined by our church/job/family/teachers/bank etc. What is so powerful about living in this time is that all of these structures are now disintegrating. We can see that the Emperor has no clothes, whether we want to or not. Some of us will have no clue as to why this is happening and for them this will be a very frightening and painful period. For others this is a liberating experience. For all of us, it is a time to be our authentic self. It is time to get into our Genius Groove!

The Call

The Call can persist for many years - usually it comes in waves over us. It can make us excited - it can make us scared. But it always brings us back to our authentic selves. As a doctor, I heard the innermost thoughts of my patients as they gave up on their dreams in order to become 'normal' - though none of them knew what that was. Who exactly are we trying to please? Who is it that really benefits if we ignore The Call? People around you may be displeased at first, but if you stay authentic, it always works out for the best. So why do we tend to ignore The Call?

Dr Clarissa Pinkola Estes in her seminal book, *Women who Run with the Wolves* speaks about 'blowing over the bones'.[1] As human beings not in our authentic selves, that's what we became - just bones. The Call, that we sometimes wish would stay buried, is The Call to blow over these bones and make them whole again with flesh: to be living as our true authentic selves.

If we ignore The Call, it will grow louder until we cannot ignore it any more. Or in some it will wither and die. Those are the people who end their lives as the shells of themselves who long ago forgot what made their heart sing.

So hear The Call, listen to it. Sing with it. Dance your raw, wild dance. Feel the shivers of ecstasy move all over you as your soul realizes that you are not ignoring the pull of your heart any more.

Sign 'o' the times

As your soul starts to rise up again, as you start to pay attention to it, your universe starts to prick up its ears. You will start to get little signs - go here, go there, do this, talk to this stranger. Our culture has taught us to ignore these signs as meaningless coincidences and not see them for what they are. They represent the magic of the universe whispering to you.

So when the universe gives you these little trails, take note as they are going to lead you closer to your authentic self, by giving you the next clue and the next. After a while, your trust in these clues will get stronger. Don't expect the universe to behave in straightforward manner though - sometimes it comes back to you in ways that your ego does not expect, but always the effort you put in comes back in some way eventually.

The Game

As you awaken, you are starting to glimpse a side of life that has not been revealed to you before. It is then that you start to slip between the cracks of the constructs that society has decided that you live in. By becoming more authentic in yourself, you can start to see through the paradigm that we are living in, or what is left of it, and view it as the false construct that it is. And even see how it has perfectly served us, but now we are moving into a different reality.

It can be helpful to also acknowledge that the banking system is a deliberate plan to keep people in poverty and that our educational systems are designed to foster conformity and take the locus of control from within your own being to outside, by putting you through examinations and tests that teach you that your value is defined by something outside. It really helps to stop watching the news, which is a device to keep you hooked into the fear paradigm.

Also celebrity culture is really prevalent in this - it creates a yearning to acquire expensive products to live an extravagant lifestyle, as if this is the ultimate way of living that we should aspire to. We are not encouraged to wonder whether these things are really important, nor to be authentic to ourselves.

Rat Race

When you move into a more authentic life, your entire being starts to change and the patterns that were right for you before, do not work any longer. This will even affect your diet. Food, caffeine, alcohol and any substances we consume are an exact reflection of our emotional states. We spend huge amounts of time trying to change our diet without realizing that we have, in our infinite wisdom, chosen the exact food for our lifestyles and are indeed medicating ourselves with caffeine, sugar and alcohol in order to survive in our society.

Once you realize this, you can actually stop judging yourself about your choices and instead see them as another example of just how amazing you are. When you feel in a different emotional place you will treat your body differently naturally and your entire lifestyle will start to change.

There is scientific evidence to suggest that this is indeed how we behave. In studies done with rats in the late 1970s it was discovered that if you treat rats well and place them in a, 'rat heaven' known as a 'rat park', they will no longer display addictive behaviors and will reject morphine-laced water, even if they were previously addicted.[2]

When rats are placed in crowded cages, they display addictive behaviors. These studies suggest that we are actually self-medicating with whichever drug does the trick, to cope with the stresses of our lives.

I believe that we all have a sense of something more and of our multidimensional nature, but often we are not conscious of it. People can become confused as to how reconcile their true nature with their daily lives - so medicate themselves in order to suppress their true longings. Once we start becoming more authentic, there isn't such a great need to self-medicate.

It's My Life

When you start to feel the awakenings of your Genius Groove, you are being called to live your passion. But nothing in your life until now has been an accident. What Malcolm Gladwell sees as coincidences that repeatedly appear in the lives of people who are successful, are actually careful choices made by the Higher Self before birth.[3]

You chose the right parents to give you exactly the conditions and the life lessons that you needed to be your true self. The schooling you had, the qualification and the jobs you have had have not been an accident. As the true authentic self in you starts to awaken, you realize that everything in your life has actually been part of a greater plan equipping you with exactly what you need.

Sweet Dreams Are Made Of This

When anyone moves towards their dreams, life can blow up in their face leading them to conclude that it is too hard and give up. In Britain, we gleefully see public examples of this as proof that people should not even start to go for their dreams and stick their head above the parapet. If we had more of an understanding of why we encounter difficult situations, as we move towards our dreams, we may react differently to them and not be so quick to give up.

This explosion of emotions is a reflection of fears that have been hidden until now and when you make a move, your consciousness shifts and as it starts to move truly into your passion, it is as if the dirt that was on the wheel starts to be thrown off. It is like that with our emotions - at some point most of us have heard an inner voice asking, 'Who do you think you are? Why should you do that? Why should you be happy? You are not clever/beautiful/thin/rich enough. Stay small where you belong'.

When you start to move along the path to your authentic self, this freaks out that part of you that wants to keep you small. Instead of seeing this as a bad sign, see it as progress! You are shining light on the parts of you that are hidden. One of the most common signs that your shadow is coming up to be noticed is that the people around you try and drag you back to where you were.

I saw an amazing example of this as a GP when I saw a young woman who had applied for and been successful in getting a new job. This prompted her colleagues at her current workplace to rally round her, telling her she was part of the gang. They convinced her that she would be sorely missed if she left and that she should stay. So she turned down the new position, stayed where she was and ended up with problems sleeping, which is why she came to see me. I asked how her colleagues were treating her now and they had reverted back to normal. The cozy, valued feeling she had experienced temporarily in her job had gone and she was unhappy again.

For that brief moment when she was going for her dreams, the aspect of herself that felt that she did not deserve to have success rose up and was reflected in the change in her colleagues' behavior. She suddenly felt wanted and loved in her job because of their extra attention and that the job was not so bad after all.

So she turned down the new job, after which her colleagues reverted back to how they were before and she felt even more miserable and could no longer sleep; her deeper authentic self knew something was wrong. Her colleagues' resistance to her changing and following her dreams, simply reflected the resistance of her own soul, as with us all. Once we realize this, then the reflections of our own self in others and difficult situations can be observed, navigated for what they are and let go; they can even be anticipated.

It ain't what you, do it's the way that you do it.

Even the most successful people in the world experience this emotional backlash. In fact, I would go as far to say that the only difference between 'successful' and 'unsuccessful' people in terms of following their dreams, is that the people who succeed are not derailed when their emotional explosion happens around them - they feel the fear and do it anyway.

I shall never forget watching a TV program charting the launch of Per Una, the fashion label sold at Marks and Spencers, a UK retail chain. George Davies had been given the task of being the designer behind the label.[4] He had an amazing track record of creating not one but two very successful clothes stores: Next and George at ASDA. Yet, the cameras were filming him on the morning of the launch of Per Una and he was relaying his nervousness. He had launched two of the most successful brands in UK retails history yet, he was afraid, like a lot of people would be, as to the outcome of his new venture.

In our own lives, we may have the impression that successful people have no fear and that is why they are where they are. So when we feel fear, we allow it to stop us in our tracks. It is natural, whoever you are, for your emotions to rise up when you venture towards your dreams and are about to do something big on your life's path. It is part of the process. Yet even those people who we think have it all sorted, get scared. The difference between people who go for their dreams and those who don't is when they feel that fear coming up they work through it instead of giving up.

You may even, like me, have a dark night of the soul and go through tough times once you start dipping your toe into the path of your dreams. But despite the tough times, I remained determined about my dreams no matter what. There were bad moments, of course, when I felt like giving up, but I always came back to my true passion. The dark night of the soul never shakes the resolve of a true path - it merely reorganizes the universe around you to help you make it happen, even though you may not see the big picture until later.

When You Put you Heart In It

When emotional stuff does come up and it is bothering you and you are finding things difficult - work consciously to find out what is going on. When writing *Punk Science*, I used to rail against this principle, thinking that if I did something 'out there' that it would bring me success. But true success can only be brought from within. Whenever I work through some emotions, I move forward in my life. The phone will instantly ring from the BBC or something like that. I also find there is no substitute for actually being in my Genius Groove in terms of moving my life forward. This is why all those people who put off their dreams until the 'time is right', rarely get anywhere with them; you have to step into your Genius Groove for the universe to respond.

In my own life, I gradually went from a situation when I had not written for ten years and had no publisher to a finished book and a publishing deal after four years of hard work. I also moved in with a loving and supportive partner, James, in a beautiful mansion in Derbyshire. The very act of being in my Genius Groove and dealing with all the emotions in my life has moved my life forward to where it needs to be in many different aspects - both personal and professional.

When writing *Punk Science*, I noticed that every time I healed to a new level emotionally, I would be given the next step professionally. For example I found the next scientific paper, crucial to my work. It was the emotional work that moved me along in my groove. It is like that crystal again that gets polished by emotional processes, becoming progressively clearer, allowing more light to come through.

Looking outside yourself for fulfillment will never work, the true pathway is within and the examining the shadow self. For it is here that we find the light that is the true nature of universe.

This is not the same as when you try to do something and it so hard you keep getting the message that this is not the right thing. Those are signs too! The difference will come from within you. You are your own best barometer of what is right for you. When you meet with resistance, do you feel it is right to forge ahead or do you know, in your heart, that it is time to let go. Developing a strong inner voice will help you. And when you really do not know, wait and the universe will show you the way when the time is right.

As it says in the *Tao Te Ching*

"Do you have the patience to wait

till your mud settles and the water is clear?

Can you remain unmoving

till the right action arises by itself?"[5]

Is there something I should know?

Sometimes it is time to let go and life will show you the way. Before I finished training as a GP, I saw an advertisement on the notice board of my hospital training center that wanted a GP to do a few sessions a week. I rang them up immediately and told them I was going to work for them. The practice manager was highly amused, but informed me that they wanted a particular type of doctor - someone who had returned to work after parental leave; a position I wasn't in.

Nevertheless, when I qualified a few months later, I saw the job still being advertised. In their desperation to find someone, they widened the job description and this time I fitted the bill. I breezed through the job interview and started part-time work. It was rather strange working there as the doctors were always very evasive and left any uncomfortable news to the practice manager to inform me of.

One day the manager informed me that the doctors had changed my terms of employment. They had decided to scrap any sick leave, study leave or holiday pay in order to cut the cost of having me. This effectively meant slicing £5000 off my yearly salary without discussing it with me first.

This was before Christmas 2001 and over the holiday, I tried to research the British Medical Association's published rates of pay to fight my case. But despite my efforts to research the rates, nothing was working. I couldn't find any printed magazines with them in and when I tried the internet in several places to find the rates, computers kept crashing. Suddenly it dawned on me I was not meant to negotiate my pay in that job. I was done.

I went into work and asked to speak to someone about my rates. I was fobbed off with promises that they would deal with it later. I realized they were hoping that if I worked for long enough under the new conditions, I would just continue and they would postpone discussing things until I capitulated. I knew then what the universe was asking of me - I had to leave. So I finished the day's surgery and never went back. I was not meant to fight the situation and that's why all the technology malfunctioned.

Leaving the medical world for the first time in ten years allowed me to really examine who I was and led to all the changes that I have described in the previous chapters.

When is it your emotional resistance and when is it a message to let go? When you feel it inside yourself or something strongly shows you the way.

<u>Keep On Moving</u>

A lot of people's problem is that they have no idea what their passion is, it is so buried under all the conditioning. Maybe they knew what they wanted to do as a child, but now perhaps driving a fire engine no longer appeals. Most people don't know who they truly are and never get a chance to explore it.

But it always begins with just listening to and acknowledging The Call within you and all the little signs that unfold as a result. There is a quote that is often attributed to Johann Wolfgang von Goethe, but apparently was actually written by William Hutchinson Murray which says,

'Until one is committed, there is hesitancy, the chance to draw back-- Concerning all acts of initiative (and creation), there is one elementary truth that ignorance of which kills countless ideas and splendid plans: that the moment one definitely commits oneself, then Providence moves too. All sorts of things occur to help one that would never otherwise have occurred. A whole stream of events issues from the decision, raising in one's favor all manner of unforeseen incidents and meetings and material assistance, which no man could have dreamed would have come his way. Whatever you can do, or dream you can do, begin it. Boldness has genius, power, and magic in it. Begin it now.' [6]

I echo the sentiment of the above quote. It is only by starting that you truly find out what the path is, a bit like the scene in the Indiana Jones film when he steps his foot out and a bridge across a chasm magically appears.

Try things out before you fully take the plunge. Sometimes it is by trying things out that you really get to find out what you want to do, by finding that what you *thought* you wanted is not quite right. Your spirit will rail against it.

For me, having been brought up in house with two parents as doctors and becoming a doctor myself, I always thought of myself as a healer of some sort - I knew no other life. After I had left my job, I thought I wanted to be a successful bio-energy therapist. So I set myself up in early 2002 with a publicist (who I met in a very natural and flowing manner) got some very nice flyers printed and had a press launch for my healing practice. On the evening of the launch, before the guests arrived, I got an uneasy feeling that something was not right. I was about to be launching myself as a fashionable healer in a fashionable town, but my heart was not in it.

The woman who was doing the marketing for me had sent out invitations to various local people including the press. On the day, the deputy mayor of Berkhamsted was the only dignitary to arrive. The rest of the guests were all my friends who were only too happy to tuck into a free dinner! I did a little speech during the evening and to my surprise found myself describing my healing practice but then saying, "What I am *really* interested in is the science behind healing such as in this book" and held up the book called, *The Field* by Lynne McTaggart.[7]

I had to do what was not quite right for me to see what was right. It forced me to look deeper inside myself. The Monday after the launch, my friend, who ran a clothes shop in town, called me. She said that someone had been in the shop who had heard all about my 'weird event'. Apparently the Mayor and his wife had informed people about my launch over the weekend and were not too complimentary. Although they were accepting and encouraging of therapies such as osteopathy and homeopathy, they had seen my demonstration of bio-energy, which involved me waving my hands about and feeling their energy fields, as decidedly whacky. My friend told me that everybody in Berkhamsted now thought I was rather weird!

The news hit me like a bombshell that Monday morning and I was distraught. I assumed that the circles that the Mayor moved included a lot of potential clients who would now be put off by his reports. I though that my career was already over before it had begun. Feeling miserable, I went for a walk by the canal with my dog, Charlie and looked at the water wanting to jump in and drown.

Suddenly I noticed through my misery that Charlie had disappeared. I found him playing with another Golden Retriever. A narrow canal tow path simply isn't big enough for a couple of boisterous dogs and pretty soon, the other dog slipped into the canal. Before she was fully submerged, I quickly caught her and pulled her out so her body was only half wet. The ladies who were with her were amazed and asked me if I was a doctor because, they said, only a doctor could have reacted that quickly to an emergency.

Now it was my turn to be astounded with their accuracy. The ladies then turned to each other and discussed how they knew the incident would happen before it did and how they were 'only just talking about being psychic'. I decided to speak up about having the same interests and we began talking. They told me about a new crystal shop that had opened nearby. I had never been to that town before, but I visited the next day to find it was only a twenty minute journey.

Within a few minutes of my entering Gaea Cystals in Wendover, I had booked the venue for my first ever science workshop and that led to another and another. So I owe my new career to trying the wrong thing first, realizing it wasn't actually what I wanted and getting so disheartened that I gave up and let go. Through the wisdom of the animal teachers, Charlie and Holly, I was guided to the next door opening. Sometimes you have to try something that isn't quite right in order to see it, but you must just try it. Step out onto your path - there are no mistakes, only things to learn.

The Never Ending Story

The truth is that your Genius Groove is forever evolving. Once you find your path, you realize you were always on it and everything in your life has led you to this moment. But then you don't stay static, by working and perfecting your art and growing all the time, you move forward even in those timeless moments. You learn more deeply how you are totally connected to your universe.

Once you are are aware of your groove, everything slots into place around it. It is what you look forward to waking up and doing. You don't feel like it is work but play. You know it is your Genius Groove because you would do this no matter what, even if you had plenty of money or if you had nothing at all.

You are in your true Genius Groove when you have become aware of your emotional reasons behind your groove and you still are moved to do it. So if, after looking at human rights issues and examining your emotions to see where you have behaved the same way either in this life or past lives, you still want to work in human rights, then it is your true groove.

But with this awareness you will begin working from a place of non-attachment in which you are not trying to fix or save anything or anybody, you just want to do it. You are doing it because the very process is joyful and makes you feel alive. For example, if someone says they want to save children in Africa it is probably because they had a bad childhood and want to save themselves. If they are aware of their emotional motives, but still want to save children in Africa because it makes them feel alive, brings them into alignment and helps them access higher aspects of the universe, then this is their groove and in a way they have set up their childhood emotional charge in order to drive them to it so they can move beyond it.

So why am I writing this book? Haven't I got some emotional charge? Am I not trying to save people by helping them to find their Genius Groove? Yes indeed. In a way this book has been a way to exorcise my demons from my life and get them all out in the open. It has been an emotional exploration and journey. But it seems to me that there is also a transcendent motive that I keep coming back to - sheer joy! To share my journey of how I found my Genius Groove and what I learnt from that in the form of a book has, for me, been about moving deeper into my Genius Groove.

I believe that The Genius Groove is our human heritage and as the conditions on this planet shift, we are being called more and more to be in our authentic selves. When you free up your emotions, your fears and your conditioning and get into touch with your authentic self - a new life will unfold for you. I call it Living New Paradigm.

Chapter 17

Shaking the Tree

"You don't need money, don't take fame

Don't need no credit card to ride this train

It's strong and it's sudden and it's cruel sometimes

But it might just save your life" [r]

Huey Lewis And The News

A New Day

It may amaze you, but myself and numerous other people have been discussing the current financial crash for about six years. We knew it was going to happen, we just were not exactly sure of when. We have been patiently waiting in the wings for a new world to unfold, as the rug is being pulled from under the current paradigm.

A few of us have just had a bit of a head start into the New Paradigm. I like how spiritual teacher, Christopher Sell puts it that some of us have woken up a bit earlier, gone downstairs and put the kettle on ready to make a cup of tea for everyone else.[1] There is no judgement in people waking up earlier or later, it just sorts out who makes the cuppa!

I have now shared with you the story of my journey from the old paradigm to the new. Imagine my shock when I read Elizabeth Gilbert's book, *Eat, Pray, Love*, and found she was also dissolving her life at around the same time.[2] As were some of my friends such as Jenna, whom I met in New Mexico in 2001 and who has helped me through some of the most difficult times of my life, as she understood the vision of this New Paradigm.

We all felt The Call. In the case of Gilbert as with me, she was unhappy with her marriage for no definite reason. In Gilbert's story The Call just kept getting louder leading to nights howling with misery on the bathroom floor. Like myself, Gilbert had a lover with an intense connection who helped to shift her out of her old life, but this relationship did not last.

This is where we diverge somewhat. Gilbert lost her house and husband as I did, but unlike me, she was given an enormous advance for the travel book that became, *Eat, Pray Love* which I suspect has more than paid back her advance by now.

You Get What You Need

What Gilbert and I do have in common is that we both were financially supported to allow us to to follow our true paths. When you truly hear The Call and commit on your path, somehow you are supported in unexpected ways and everyone's journey is different. I used to be a self-confessed control freak; I liked to know where I stood at all times and planned things very carefully. I was always living from a paradigm of fear, worrying about what might happen. What I feared most is not having a salary at the end of the month. I lived for that paycheck and believed it was my lifeline.

I could not imagine what it would be like to not earn and bring home the bacon (as well as the tofu). So life had to ease me out of this in stages. First there was that incident when the psychiatrist turned up to my home and I was put on sick leave. It was a huge loss of identity, but I still had my job at the Bristol

Cancer Help Center so it wasn't a such a big wrench out of system at that point. I also believed that I would be reinstated by the GMC soon.

But even in those dark days, I had what I needed to get by financially which just astonished me. I had the money to pay the last bit of rent before moving back to my own home be-cause I had done a talk in Ireland and had created just enough Euros to pay it. I found I had just enough money to fill my car with petrol to get to Bristol or to talks, even if I did have to plan ahead by a few months. My ex-husband was paying the mort-gage and the household bills, but I still had to take care of my personal bills and somehow I would get the money. When I needed something special, such as the money to go to a confer-ence, I would get a few more bio-energy clients or money would come from some other source.

What really amazed me, was the kindness of people, even people I barely knew. Suddenly people would give me things: presents, taking me out for meals. Presents just started to ap-pear in my PO Box sometimes from complete strangers and out of the blue. They were little tokens, but just meant so much to me.

My universe seemed to be showing me just how abundant it really is. I discovered to my surprise that if you follow your heart, the universe does actually support you. For the first time in ages my bank account did not go overdrawn - ever!

One of the most dramatic demonstrations I had from the universe of its abundance was when I was feeling distraught because I couldn't pay my car tax. If I couldn't pay my car tax I couldn't travel to Bristol for my last available source of income and would therefore be totally broke.

I wailed in desperation and frustration and slammed my head onto my desk in defeat. My office was on the top floor of my Berkhamsted home which was a three-story town house. Suddenly a strange things happened, a robin appeared on the top floor outside my office - actually inside the house. I felt a wave of energy over me. Of course my dog, Charlie started to have fun trying to catch the bird, so I helped it to leave the house.

It seemed so odd that in my moment of desperately crying out to the universe that this red breasted robin appeared. It felt like a message from the gods! I picked up the phone to Bristol and asked the accounting department if they had paid me for previous work as they had told me there would be a delay. A slightly annoyed voice in the accounting department informed me that I had already been paid the day before and if I had bothered to check my account, I would have seen that £800 had been paid to me.

I put the phone down not mentioning that they had over-paid me by £300! Somehow I had enough money to get by *again*! It felt as if my prayer had been answered. These types of incidents kept happening until I realized finally that the universe *is* abundant. Like the song goes, you may not get what you want, but you get what you need.

Kings of the Wild Frontier

Now I am not at all advocating giving up your job, home, relationship all at once to see if the universe gives you a lottery win. You still have to meet the universe half way and do your bit. Ignoring the dimension that we live in is not the aim of spiritual growth, although a lot of people think it is. It is about aligning your life in this dimension with that of your Higher Self. This occurs through the process of surrender.

After quite a while, my ego stopped fighting so strongly against the changes. This was difficult due to my control freak nature, but I was starting to enter the state of surrender.

A number of my friends also had heard The Call at around the same time and had the inexplicable desire to just give up their old life and allow the new to come in. We started living in a kind of parallel universe next to the world of jobs and structure and somehow we were feeding ourselves and had everything that we needed in various ways that were right for each of us.

I suddenly had to face what many people are facing now - a loss of identity. I was no longer working as a doctor, but I had nothing else to identify myself with. I was writing, but I had no book and no publisher. I dreaded people asking me what I did. If I told people I was writing, I would invariably get asked how I made a living.

I needed to learn that I was not a doctor, a wife, a daughter, a homeowner or any of these things. There was something so much deeper in me that needed to come out. Gradually I have been rebuilding myself up and entering into a new world and I can choose to take back into it what feels right.

What I found was that I was living a totally new way - free, but strange at the same time and I could not go back. I call this New Paradigm Living. It is hard to describe New Paradigm Living - to describe it would be to understand it and to understand something it needs to have a definition and structure and that is the whole point of the New Paradigm - it flirts with structure but doesn't ever embrace it. Nevertheless, I will try and describe some the characteristics of the New Paradigm lifestyle.

Don't Push It

I was taught, like all good little career girls in Thatcher's Britain, that you could have it all if you went for it, you had ambition and worked hard. My idols in the 1980s were women who were at the top of their careers and displayed these values. Until 2002, I still believed that anything was possible and I could achieve whatever I put my mind to.

However, in my new life based in the New Paradigm, this just doesn't work anymore. After I lost my medical career, I tried to apply for other jobs. This used to be quite a simple procedure involving filling in application forms and sending off CVs/resumes before being shortlisted for an interview.

I soon found out that this is not how things worked anymore. Everything that could go wrong in the process, seemed to do just that. Emails and internet would just refuse to work. Phones would cut off. I started to get the message that I am just not supposed to do this, I am not meant to push to make something happen.

One of the biggest lessons lay in finding a publisher. In my old world if you tried hard, you succeeded, but not in this world. No matter how many publishers I applied to, not one took it up. A particularly bad knock-back in 2004 left me very distraught and almost wanting to give up. Yet I didn't realize that I was being given a lesson in timing. When the book was ready, the publisher appeared. It was not a case of me pushing to make it happen as I had always been taught. I had to realize that everything has its own timing and no amount of pushing is going to make it happen.

Nor is it a case of just lying back and letting the 'angels' do all the work for you. You still have to show up: make the calls that you are prompted to do by your inner callings or get in the car - angels can't drive your car for you - well not always! It is not a case of ignoring this dimension. You also need to listen to your inner voice.

Many people find it very difficult to hear their inner voice - spiritual practices may help this, but if you are really walking your path and have followed your inner voice, it develops strongly like a muscle. You start to sense an effortless flow and a rhythm in your life as you follow your inner voice or are shown the way.

Relax

Sometimes it is not as easy to make a decision, but usually time shows you. Or if you really cannot feel either way, you can relax in knowing that all you need to do is choose and it will always be the right path there can never be any wrong decisions.

One of the most important principles of New Paradigm Living is the future has already happened so every decision is the right one. In the post Law of Attraction world, instead of feeling liberated, some people are feeling stressed out by the pressure to manifest. Although making the connection between our outer and inner lives is truly a revolution, sometimes manifesting does not always seem to work.

When you realize that the future has already happened, you can relax and flow with the choices you have already made for yourself. There are absolutely no regrets as you know that each decision is the right one and that there are no missed opportunities. You are always in the right place at the right time.

I have taught this principle to hundreds of people over the years and they all suddenly feel a huge burden lift as they understand the implications. It is the one major message you can take away from this book that will instantly make a huge difference to your life - relax - it's already happened!

Traveling Down My Own Road

When I met Professor Richard Dawkins whilst filming a TV documentary, he spent a lot of time staring at me curiously as if he had never met such a creature. One of the things he asked me off camera was what exactly I did all day? To be quite honest I don't know how my days fill up, but they do. Even if my diary looks empty at the start of the day, things tend to happen like the phone will ring or there will be an email to show me the way to move forward. Or I will receive promptings within me to take some sort of action. I then have to do the work of following the lead or getting ready to act on the prompting. The days of planning everything down to the last minute are over.

An article in a 2002 edition of *Fortune* magazine, followed the business practices of Oprah Winfrey.[3] Here was one of the richest women in the world being interviewed by one of the foremost business magazines and of course *Fortune* wanted to know how she does her business planning and strategizing. At this point Oprah and her staff fell about laughing.

The truth is they have never held a strategy meeting and go as far as saying that the staff would think it was a joke if someone suggested it. Isn't that extraordinary? Winfrey has created fabulous wealth with no formal business planning. She has indeed made some high profile course corrections over her career, but she has learnt from these and shares her lessons quite openly.

When you live New Paradigm you realize that you have surrendered to what your Higher Self has in mind for you. Even though your ego wants to keep hold of the reigns, after a while you realize that your Higher Self knows much more about yourself than your ego does. I often say,"Oh, this is what I am doing now apparently" as I see a path unfolding in the moment. I have long given up the reigns.

For me I can tell when something is not right for me - I get a strange feeling about it. This may seem strange to those around me. On occasion, someone may suggest to do certain things in my business, but if I feel it isn't right for me, I will turn it down, giving no other excuse other than I just know. But I am not someone who just sits back and lets things happen - a total lack of action is not right for me either - we each have to find the balance within ourselves, just as the Black Hole Principle involves both matter and antimatter as well as light and flows between the two states - the universe is in constant flux between heaven and earth.

Nothing Really Matters

When you are really in your Genius Groove and in your flow, you are not working from your ego. Therefore, when you receive praise from someone, it does not inflate your ego as you are not working from that place. So you will not be boosted up, only to drop five minutes later.

You also do not get deeply affected by criticism as you are working from your deep core. You understand that these are part of the balance of life. You are centered in yourself and no matter what praise or criticism comes your way, you would do what you are doing anyway as it has never been about the external - it is about the groove itself - that is its own purpose.

I saw an amazing example of someone who was totally in their groove and oblivious to all else when I saw chef, Gordon Ramsay being interviewed on the Jonathan Ross show.[4] I was out in Sussex in a tiny hotel room, due to give a talk the next day. There was not much to do other than watch the TV and when I switched it on, I heard The Genius Groove in action.

There was Gordon Ramsay who is notorious for hurling his bad temper at his subordinate chefs when he feels they are not performing. He is also a brilliant artist with a global reputation. But what Jonathan Ross was interested in that night was finding out what issue really annoys Ramsay and causes him to fly off the handle. It was obvious that Ross was expecting a salacious answer about people messing up in Ramsay's kitchen or something similar. The answer was even more shocking; what really bothers Gordon Ramsay more than anything else in the world is seasoning.

"Seasoning!" Ross repeated incredulously and the audience started laughing thinking it might be a joke. Ramsay looked up at them, slightly bewildered and at that moment I knew I was looking at the face of a genius in his groove. Famous as he is, bad tempered as he may be, none of that really matters to him. What matters is his art and although that was what *he* cared about, he didn't realize how others saw him. What bothered Ramsay most in the world was when people didn't add seasoning early enough in their dishes - the answer of a person truly in the groove.

Perfect Day

The art of manifestation is useful, but after you have learnt that lesson you will move on and start to take yourself into non-judgment and the surrender process. You realize that as my friend, Story Waters says, *You are God, Get over it!* [5] Paradoxically if you are living in the flow - there will be times when you are called to manifest something. That too, is just part of your flow.

So don't even judge manifestation as there will be times when that is the right thing to actually do! The natural change in vibration that happens as you go deeper into your authentic self will start to manifest your true world around you, but the changes are from within.

Mirror, Mirror

Not only are you are always in the right place at the right time. You are attracting what is perfect to you at all times: the life that is a reflection of you at some level. In this way you are living the oneness - whether you like it or not. So there really is no judgement to anything or anyone - it is all a reflection of you. If something makes you feel uncomfortable then you have to look within you to see why and work out what emotions are being triggered.

Debbie Ford has written some excellent books with many case studies demonstrating how to uncover your shadow emotions.[6,7] Sometimes people just cannot see how they are reflected by what is bothering them, but that is the point - it does take time on our journey to learn about ourselves. It really is the secret to freeing your life and your relationships with everything and everyone. This doesn't mean that you will end up chanting in a serene cloud of incense, but if you really do keep the awareness that everything is a reflection of you, things don't bother you like before; you understand that everything is a message to you, from you.

I notice how little I get angry nowadays (although I definitely still do). If anybody starts to shout at me - I start to think in the moment as to what it is all about. Usually the anger of the other person dissolves as soon as I do this inner work. So the New Paradigm is that you live as part of the oneness - you are connected to everything and everything is a reflection of you at some level - even the aspects of life you don't like. I promise you, if you start to live like this, it will make a huge difference to your life and relationships.

Break Every Rule

If we knew everything about our lives and all of the rules we wouldn't be learning anything would we? As it is with New Paradigm - I live everyday with no idea at what it is all about and what I need to learn next. As Socrates is quoted as saying, 'all I know is that I know nothing!' And this is why I wrote this book. There was no urge from my ego, there was no urgent financial need in my life. From out of the seed that I had in me, a book has grown and the book has surprised me.

As I started writing, signs appeared around me. It has taught me that I really don't know how things work as new things happen all the time for unexpected reasons and there is always something different to learn. I hope you enjoy your own Genius Groove and your journey into the New Paradigm. And remember -

Your Life just Got Groovy!

Part Eight - One Nation Under a Groove: The Genius Groove for Organizations

Chapter 18

New Power Generation

"Win or lose, sink or swim

One thing is certain we'll never give in

Side by side, hand in hand

We all stand together" [s]

Paul Mc Cartney

What if throwing everything out of your life is not for you? What if your Genius Groove does not involve being an entrepreneur and going it alone? Does living New Paradigm mean that everyone has to give up their jobs and their former lives? What if it is right for the call of your heart to work within an organization? How do you live your Genius Groove and your creativity and still keep your job?

As Richard Barrett has pointed out, 'the successful organization of the 21st century is the one that has the ability to turn any job into a mission and make 'personal fulfillment paramount'.[1]

But this is far from the daily reality for many people. One of the reasons why people often feel their jobs do not fit them any more is that organizations tend to follow rules that fit the old paradigm. This sees the world and the people in it as soul-less machines, separate from each other.

However, there is a movement that is gradually gaining ground in organizations and is starting to change the way they are run. This movement is examining the implications of the New Paradigm based on the New Science and how these ideas can transform our organizations and the way in which we work.

The financial crisis that unfolded in 2008 has only served to emphasize that 'business as usual' cannot continue. This is a time of tremendous opportunity to make changes within our structures to bring them into line with the science and philosophies of the new era.

In this segment, we shall explore how some of the ideas in this book and in the New Science are starting to transform organizations. I shall be using evidence and quotes from some of leading business visionaries to illustrate the changes, particularly those from

- Margaret Wheatley in *Leadership and the New Science*.[2]

- Bill Guillory in *Spirituality in the Workplace*.[3]

- Richard Barret in *Liberating the Corporate Soul*.[4]

- C. Otto Scharmer in *Theory U* and *Presencing* [5,6]

- Ricardo Semler in *Maverick*.[7]

The 'In' crowd

Human organizations, whatever their size or stated purpose, all have some things in common - they consist of a group of people relating to each other and their environment. This group of people come together to interact with each other in order to impact their universe. These same principles apply whether the organization is the local bridge club, a small business or a huge multinational company.

At some time or another, we have all experienced what it is like to be part of a group. The vast majority of us have also experienced some of the pitfalls, for example - countless meetings in which the people who shout the loudest seem to get their way. As people's consciousness and awareness change, these types of interactions are no longer enough - there is a deep yearning to become more authentic.

The change in consciousness within people is occurring on a global basis and in order to keep up, our organizations and the way in which we interact with each other also need to change. These inner changes are linked to the radical shifts in the way in which we understand our universe, as described throughout this book, with the New Science of quantum physics, the New Biology and even the Black Hole Principle. This New Paradigm requires new ways for our structures to operate. However, the vast majority of our organizations still operate under the old rules - hence the frustrations. What will happen to the way in which we structure our organizations, relate to each other and our universe once we take these new scientific ideas on board?

Above is just a sample of the pioneering authors and visionaries who are already asking this question. They are taking these pioneering ideas from the New Science to within organizations and Universities such as MIT.[8]

This movement is constantly gaining momentum, with companies like Change the World Corp also picking up the baton.[9] We are headed for a powerful alignment between the worlds of cutting-edge science and business, which is going to radically change our organizations of the future.

Until now, although our daily lives are filled with new technologies, actual cutting-edge scientific philosophies rarely have had an impact on our lives. The revolutions that have rocked science in the twentieth century, such as quantum physics, have generally not filtered into the rest of society.

Now these worlds are converging. The findings and implications of quantum physics are being embraced as principles that can be applied to our lives. This is happening both in organizations and in the personal lives of millions of people. Although it is hard to give equivocal numbers on this change, there are some important indicators, which include the explosion of book sales in a new genre - those that bring together ideas in quantum physics with applications for everyday life. And there is one book in particular that stands out as an example of this movement.

Secret Garden

I picked up the Sunday Times bestsellers list for the UK this week.[10] The book that has spent by far the longest time on the bestseller list is *The Secret* by Rhonda Byrne with 85 weeks in the top ten as compared to 47 weeks for a book on the Eden project, the next longest running book.[11]

That's more than Jade Goody's diary (15 weeks). That's more than Jeremy Clarkson's latest offering (11 weeks). In fact subjects we give attention to in our media - reality TV stars and fast cars, even eco gardening, actually pale in comparison to the subject many are choosing to learn about for themselves. *The Secret* really amounts to a shift in an understanding of reality.

The Secret is discussing the Law of Attraction, whereby our thoughts attract situations in our life, as described in previous chapters of this book. It makes use of the findings of quantum physics to make the connection between our consciousness and the world that we see around us. On sheer volume of sales, *The Secret* is obviously a social phenomena - despite the fact that the UK mainstream media don't seem to be noticing. A revolution is going on under our noses and the only mention in *The Sunday Times* is that it is number five in the 'Guides and Manuals' chart. Despite the media silence, the New Science is becoming mainstream and we are beginning to incorporate these ideas into our lives.

The high sales of *The Secret* is just the start. The links between quantum theory, the science of consciousness, chaos theory and our working lives has been made by visionaries such as Margaret Wheatley in her ground-breaking book, *Leadership and the New Science*.[12] She and others have done much to pioneer the New Science into organizations and explore how these ideas can change how we live and work.

The last decade has also seen enormous changes in cosmology giving us the concept of fractal cosmology and a new vision of black holes. We have gained even more of a picture as to how our universe works, which can then be consciously applied to our daily living.

We now have an amazing opportunity to create a new way of working with each other and relating to our universe. So let's start looking at some of these principles and how we are leaving the old era behind and entering a new one.

Rage Against the Machine

As mentioned previously, one of the major pioneers of the movement to bring the New Science into organizations is Margaret Wheatley. In the 90s, she could see that changes in physics, as famously documented by writers such as Fritjof Capra, were charting a passage of a social revolution from the old mechanistic paradigm to the weird new world of quantum rules and beyond.[13,14]

Organizations are often still stuck in the mechanical model of the universe in which humans are separate from each other and their universe and are passive observers of their world. The idea that we are machines in a mechanical world has been an integral part of the thought patterns behind organizations and indeed could have been an important aspect of the industrial age, when factory workers were seen as merely cogs in a greater system.

With the advent of quantum physics, we realized that we are not separate from our universe as the connection was made between our observation of an experiment and its outcome. We realized that it is our participation that brings reality into being and collapses the wave into a particle within the atom.

Some have concluded that physical reality is a outcome of consciousness and not the other way round. Consciousness is not just a side effect of the workings of the brain, but is the elusive factor that lights up neurons in the first place.

The effects of the changes in science are no less than revolutionary on our world. Their implication is that we are not just soul-less cogs in the wheel, but that even our thoughts have an effect on our universe and not just our actions. The concept of the machine universe, populated by humans who are not much more than molecular robots who happen to have thoughts, is on the way out. Let's have a look at some of the implications of this shift starting with the concept of a participatory universe.

All Hands on Deck

As science makes this shift from a world in which we are passive observers to one in which we are participators in our reality, analogous ideas are starting to take hold within organizations. In fact, it is one of the key concepts listed by business visionary and author and founder of the Values Centre, Richard Barrett in his book - *Liberating the Corporate Soul*. [15] He speaks of the importance of a sense of shared ownership for the well being of everyone who works within an organization.

A well-known example of a company that has moved from a traditional mechanical, hierarchical structure to one whose business model is all about participation and taking ownership, is Semco in Brazil. The spirit of democracy is so strong within this company that the factory workers even decide their own salaries!

The story of the shift from a traditional to a forward-thinking manufacturing company has been documented in a million-selling book, *Maverick* by Ricardo Semler.[16] It was first published in 1993 and became a global success. However, there were also a lot of detractors who felt that such a non-hierarchical system cannot work in practice. As Semler points out in the introduction to the latest edition in 2001, these detractors are now long gone, but Semco remains.

The democratic model that Semco exemplifies is becoming more influential as we move deeper into the new millennium; there has been an explosion of interest in social enterprise. These are companies which not only incorporate a social purpose, but also include a democratic model of management with 'members' each having a vote in decisions. Some major companies in the UK such as the John Lewis group are run according to the social enterprise model.[17] There has been a huge surge in interest in social enterprise over the past few years as more people want their work to make a difference as well as be more democratic and participatory.[18] This is all part of a movement from strict hierarchy to shared control within our organizations.

Control

There are many psychological studies which show that a lack of control of your environment causes stress. [19] If people feel powerless to speak about the way in which their organization is run, they simply shut down and no longer engage in their creativity. This powerlessness, when sustained over long periods of time, leads to the type of disconnection that is common within our society. Hence we have a culture of 'keep your head down' and 'job's worth' automatons when we could be tapping into the infinite creativity and Genius Groove that lies in the heart of each person.

In the new scientific paradigm we are not just soul-less cogs in a machine - we are multidimensional beings, participating in a universe alive with consciousness. As we make the transition into the New Paradigm of a participatory universe, models of involvement within organizations, such as those in social enterprises and those demonstrated within Semco, will become more prominent.

We are also starting to realize that because consciousness is fundamental to reality, the material world is formed according to its patterns. Whereas the old paradigm only considered the physical, material world, in the new era thoughts and consciousness are seen as fundamental. One of the key implications of this realization is that we can reconsider the impact that collective intent has on an organization.

Together in Electric Dreams

Have you ever walked into a workplace where people are happy and productive? What about when people are unhappy? Even if nothing is being said, you can pick up a prevailing mood from the atmosphere in a place. Before the era of the New Science, such ideas were seen as superstitious hunches, but now we can explain them scientifically.

We have actual evidence that people's repeated thoughts can contribute towards the atmosphere of a place. If people in an organization are unhappy and are continuously having thoughts against their environment, the scientific evidence suggests that the space actually changes to reflect these thoughts.

An intention can be defined as a focus of consciousness towards an outcome. When an intention is repeated over time, it will even alter the space in which the intention is made. Whatever you are intending for also becomes more likely. So over time an intention that is repeated in a space, even if it is not verbalized, is able to affect an organization.

The power of intention has now been studied under scientific conditions around the world and many of these studies demonstrate a scientifically significant effect that appears when a person, or group of people, focuses on an outcome with a specific intent to influence it. In *Punk Science*, I suggested the power of intention works through an aspect of quantum physics called non-locality whereby two particles are connected no matter the distance between them.[20]

One of the research teams studying these effects includes emeritus Professor William A Tiller of Stanford University. His team's research on intention are published in books such as *Conscious Acts of Creation*.[21] Whilst performing his experiments, he realized that certain aspects of the space in which the experiments were being performed were being changed by the experiments themselves.

As time went on, the space continued to change in quality and concurrently the success of the experiments increased. The work of Tiller and his team opened up the possibility that repeated intentions actually change the space you are in. This has huge implications for the workplace.

It is not just our actions that influence our world, it is our repeated thoughts too. Some companies implement policies whereby people are not allowed to speak badly of the organization. But if the people in the company are not happy, the problems have not been solved. Their very thoughts will influence the atmosphere and condition the space to make certain outcomes more likely. Because the universe is holographic and fractal, one aspect of consciousness affects the whole.

So the power of consciousness needs to be more deeply understood as a hidden force in our organizations so that problems are not just swept under the carpet. Conversely a positive outlook should create positive intentions and an upward spiral in terms of outcome. But does all this really matter if, according to the Black Hole Principle, the future has already happened. Does it matter what our intentions are if the outcome is going to be the same anyway?

I've seen the future and it will be

If the future has already happened, the outcome of any intention has already occurred. However, as described in previous chapters, intention is a powerful stage in making the connection between consciousness and physical reality, before the stage of surrender.

So the science of intention can be a useful tool in leadership and may be the key to transformation. This transformation has already happened, but using this knowledge about the science of intention may bring about the changes that were already destined to happen as causality is circular. (You may need to read this last bit again!)

Both Otto Scharmer and Joseph Jaworski speak of leading from the future as it emerges.[22, 23] This is the space where you can actually sense the future and what needs to happen rather than forcing your ego's will onto the universe which was the old 'push push' way of accomplishing things.

They speak of 'presencing' - the place you reach within when you are in direct relationship with your universe - when you are listening to the space and what needs to be brought forward as a group. Often we are too busy pushing our ego's will onto others to actually be listening and partnering with our universe. Scharmer speaks of going through the 'letting go' process, including facing your emotions which he names as the voice of fear, the voice of judgement and the voice of cynicism.[24]

Once someone has overcome these barriers, they can truly connect to that authentic space within and be able to deeply listen to others and to the universe and to what needs to emerge in the moment. To do this, you also have to be able to sit quietly with yourself.

A lot of people are in too much emotional pain to be comfortable with stillness, so they busy themselves all the time as they are afraid at what will emerge from the recesses of their psyche if they stop. We also spend so much time in the same old routines, not stopping to see what is needed in the present.

In the next chapter we shall be discussing some more about emotions and how to enter that place of comfort, but for now, let's examine the creative zone of groups or the collective Genius Groove.

Shapes that go together

This state of connection or 'presencing' that Scharmer discusses is the same as The Genius Groove. Is it possible to reach this place within a group? When people are able to gather together having let go of past patterns and feel safe with each other, they can enter a zone in which they can access a collective Genius Groove. By accessing the Quantum Vacuum as a group, information emerges that is actually greater than what was possible by the individuals working on their own.

Those moments of collective flow are what everybody is really looking for when working in a team, but sometimes this space is not reached. Sometimes, it can be due to just one person operating from the space of the old paradigm, being driven by ego drives for dominance and power. This is also an opportunity, of course, for the rest of the group to see what is being reflected in them.

The scenario is a familiar one to many, as meetings tend to end up with a few dominant voices. There are many such procedures that are often dissatisfying. There is a chance now to understand how to access an authentic space on a collective basis and allow true group creativity to emerge and we are just starting to see the scientific evidence of this process.

The Heart Math Institute have studied what happens when people connect with each other in a meaningful way and have found that their heart waves actually come into sync for some time after they have stopped interacting.[25] It seems that when people have a meaningful interaction there is an effect that is measurable - a sort of group mind.

Tiller has described how, when two people are engaged in a meaningful interaction, there is not only a synching of their heartbeats, but the capacity of the system to take in information from the universe increases - in this case the system is a couple of people having a meaningful interaction.[26]

Although the data is more sparse for larger groups of people, the evidence suggests that this is also true - that by moving into a heart-felt, authentic space, the capacity of the system to receive information increases. This means that when a group of people are truly heart-centered together, their ability to be creative increases and they are able to channel more from the Quantum Vacuum as a group than they would as individuals. So the scientific evidence of a group mind is growing and demonstrates how creativity increases when the connections between the group are more authentic and meaningful.

The space of flow/presencing or collective Genius Groove is the real key to creativity in an organization. The steps to get into this space may sometimes require some emotional resolution and seem difficult at first, but lead to the true source of collective creativity.

The stages of getting to that space can be better understood by using the scientific principles described in this book. For example, emotional resolution has been described in terms of collapsing antimatter and matter in light through awareness, creativity has been described in terms of bringing light from higher dimensions via the Black Hole Principle. Understanding the science of creativity and the dynamics of group mind, may help people to reach the space of the collective Genius Groove more easily.

Roll with it, Baby

There is much emphasis from business leaders on the importance of managing change to the longevity of any company. However in practice, we are often under the illusion that it is possible to gain stability and keep situations fixed as they are. But to fix something down so it cannot move is akin to death and this way of thinking belongs to the old paradigm of the universe as a clockwork machine.

The new model of the universe with its breathing, flowing processes in everything from the atom to a supermassive black hole, helps us to understand that a balance between two aspects of the universe is essential to survival.

As we have seen from the Black Hole Principle, there is no set point to the universe; nothing stays stuck and there is constant movement between the matter/antimatter and the gamma ray/light aspects of the universe. If you try and fix something and hold onto structure too tightly - something Margaret Wheatley calls the 'death watch', it is like trying to keep something in the matter and antimatter region of the universe.[27]

This is what happens when an organization gets bogged down with structure and becomes less able to move. This situation can happen if the mind set of the people running the organization is based on fear with people wanting to hold on to the ways of the past. The antimatter/matter region of the organization increases just as in a person's emotions can get stuck in the space and time when they experienced a trauma and end up getting cancer, a process discovered by doctors such as Ryke Geerd Hamer.[28]

When a company gets bogged down in procedure and structure, it is as if it gets cancer - it has too much 'stuff'. There is not enough returning to source and the light, leading to a loss of reflexivity and creativity.

Conversely, an organization that stays in the light region of the universe at the expense of the antimatter and matter regions will never get going. With inadequate structure, it remains ungrounded and will dissipate before being fully actualized. What is needed is a balance of the two. A truly conscious organization can take the lessons of the Black Hole Principle and allow itself to breath between the two aspects of reality.

I was amazed to find references to leadership as a breathing process appear in the works of visionary business thinkers. Scharmer speaks of leadership as an 'in and out breath'.[29] When you breath in, you breath in to source - this is the process of letting go of your familiar patterns and reaching that authentic space. When you breath out, you then take that source back to the world.

In the Black Hole Principle, the breath occurs at every single level of the universe in and out from the universal source of light and consciousness; the breath provides the pathway to and from the creative source. In any creative endeavor there needs to be balance between the in and out breath. Too much on the side of the light and source and an idea remains ungrounded and is never put into action.

However, if it stays on the matter and antimatter side without ever returning to source, actions becomes stuck and outdated without being periodically renewed from the source. As we allow our organizations to breathe, by letting go and being present, we can discover the inherent order that lies deep to reality.

Spiraling out of control

As we have seen, the same self-organizing pattern is inherent throughout the cosmos - that of spirals displaying phi ratio. Nature organizes itself according to spiral principles. This is also touched upon by Scharmer - that the emerging pattern of a new field is a spiral.[30]

We now have the evidence that everything from galaxies to stars, to plants to our own bodies display these spiral, phi ratio patterns. The universe consists of infinite creative spirals and is self-organizing; this is a fundamental aspect of creation. Organizational systems are no different; they will self-organize according to the underlying consciousness. The trick is to let go enough to surrender tight controls, but keep enough structure so the fluidity has some sort of container. My friend, Jenna and I often call this concept 'structureless structure'.

One of our major worries is that if we let go and surrender, chaos will ensue. But as Margaret Wheatley points out, according to the New Science, seeming chaos is actually deeply ordered.[31] As organizations like Semco have found out, if control is relaxed, people will organize themselves.[32] Semco work on an eclectic bunch of products, a policy which breaks all the received management rules, but they do so because that is how the workers themselves have organized the company. And they are still highly profitable.

But on the other hand Semler says that he spends time weeding out applicants to the company who believe that people are paid just to do nothing all day.[33] It is important to get the balance right so that true creativity can flow.

We often forget that all biological systems have the ability of homeostasis - of self-referencing and adjusting according to feedback loops. If we are able to be emotionally coherent enough to let go and surrender to what needs to happen without giving over completely to 'structurelessness' we can allow for emergence and the homeostasis of the system to take place.

Where's the Party?

It takes a shift in mindset and, most importantly, emotions to be prepared to make the changes that have been discussed in this chapter. To allow for surrender of control, fluidity, participation and to be present with the universe and others, it takes a certain amount of emotional awareness and courage. Why is that most companies seem to be unable to display such traits? Because often, by the time we are adults we have already gained layers of conditioning and emotional stories which hold us into places that we know and are familiar with - 'the way its always been done'.

However, there are also people who are willing to make that shift. This may not always be the easy path as it take courage to change, to shift frequency and to examine the shadow on an individual and collective basis. But the rewards are no less than a new dimension of living. In this next chapter we shall explore the role of emotions within organizations and how, when their dynamics are understood, they are in fact our most powerful transformational tool that will take us into a new era.

Chapter 19

My Ever Changing Moods

"And inside we're all still wet

Longing and yearning

How can I explain how I feel?

Let's get unconscious, honey" [t]

Madonna

Coming Around Again

What determines the way in which we relate to each other and our universe are our emotions. As stated earlier in this book, our emotional life is the key to understanding ourselves and our behaviors. Without understanding our subconscious emotions, we cannot comprehend why we behave in the way we do.

As discussed previously in this book, our emotions in early childhood determine a lot of how we behave as an adult. Incidents affecting the emotional body occur even as the soul is downloading into the womb, creating multidimensional geometric patterns that set up an emotional journey which plays out throughout the person's life. Ultimately this journey, has already been chosen by that soul beyond space and time.

Collectively as a species right now, our emotional patterns and behaviors are changing. This is likely to be the result of the shift in the Earth cycles that, according to many traditions around the world, last 26,000 years. There are some modern theories that link this Great Year, as it is sometimes called, with the cycle of our sun around a binary star.[1] Many ancient traditions speak of a golden era before the fall of mankind into the dark ages.[2] The dark ages represent the era of belief in separation, because it is only when we believe in the illusions of separation that we can tolerate war. We have come from a golden era into a dark era and will cycle back to a Golden Age.

This is why our collective emotional patterns are shifting, because consciousness itself is cycling. As we move into an era of enlightenment, more people are becoming spiritually conscious and we can no longer tolerate old patterns.

These awakenings are happening in all areas of life and the discontent with the status quo is being felt throughout our organizations. We are currently creating new patterns to reflect the changes in the underlying consciousness and our emotions. We can no longer tolerate the place from which we have previously been operating from.

Scharmer speaks of the Blind Spot in our current organizations.[3] We examine all aspects of life based on physical results, but usually nobody stops to consider the interior space from which we act in order to obtain the results that we are getting. As we go through these collective shifts in consciousness, our interior space i.e. our emotional patterns are changing. At the moment, we are at a cross point, where some people are making major shifts in their consciousness and behaving in a different way and others are trying to stick to the old rules. This can create a lot of confusion, unless it is understood as to what is going on.

Kid Fears

The bottom line is that it is difficult to make the shift in your emotional patterns from a controlling, hierarchical, mechanical paradigm to a open-hearted loving one if your emotional life is holding you back, for example if your needs as a child have not been met. If you feel that your nurturing has been lacking in some way, and this may be totally subconscious, as an adult you will always be trying to get your needs fulfilled. You will try and obtain what you perceive you didn't get in childhood from everyone around you and may even try and punish them for what you feel you lack, to try and heal the hurt inside. It also makes it hard to really take responsibility as you are always trying to make others responsible for your pain.

Effectively, a lot of adults are still in this emotional emergency mode. Whilst in this state it is not easy to be expansive and loving. The behavior patterns of power and control over others are often displayed by those who are in the state of emotional emergency.

Within the arena of organizations, all of our emotional stories are played out. The hurts that have been unresolved with our families pop up in our interactions with colleagues. Some people may be driven to reach the top, gaining power and status in order to fill the gap that their perceived rejection and lack of love has left inside them.

Becoming aware of the subconscious patterns and resolving them, as described in previous chapters as the process of moving your awareness out of polarity, is the way to expand the heart and move into the space of unconditional love. Until a person makes the connection between the subconscious emotions driving them and their behavior, they may not have the capacity to open to new ways. Of course, this journey never stops and deeper layers are always unfolding within us.

To get to the point where you move from the power hungry based state to the heart-based state is to go 'through the eye of the needle' as Scharmer puts it.[4]

Richard Barrett speaks of the 'transformation of the heart'.[5] To heal our perceived dysfunctions of our organizations is to heal the emotions of those individuals within them and realize that they are not dysfunctions at all - simply emotional aspects being played out in an arena for learning.

Jack your Body

So how do we stop shoving around emotions in our organizations like some unwanted vegetables on a plate? How do we go through the eye of the needle and move from being in a place of fear to one of a deep, authentic relationship with the universe? Well, if it is in your soul plan to go through the eye of the needle, you will do. It takes different paths for different people - from spontaneous epiphanies to emotional healing therapies and much more. Everybody's journey is different.

But essentially the effect is the same - to move from a predominantly power based mode of operation to a predominantly heart based, loving based mode of operation, although the two are not mutually exclusive. I have adapted an idea from author and lecturer, Lucia Nella and have found this to be a good model to help people become cognizant of the different modes of operating although it is not scientifically tested. [6]

In Chapter Ten, we explored the different dimensions to our beings including the emotional body. We can think of the emotional body as needing food - it feeds off emotions and energy in order to sustain itself. Ultimately this is what all food is - taking in energy in some form. When we eat physical food, what we are actually doing, through a series of biochemical reactions, is releasing the energy from sunlight into our bodies. The ultimate fate of much of the food that we eat is to provide us with energy. You can think of our emotional bodies in a similar way - they also need to be fed with energy.

Now think of the emotional body as feeding in two different ways - horizontally or vertically. In our society we often are given lots of food for our emotional bodies to eat in a horizontal fashion. Hence we get fed by listening to other people's misfortunes - e.g. the latest celebrity crisis, the news from a war zone, horror films, or the latest gossip from our office. Because this energy is actually feeding our emotional bodies we get hungry for it, if it has been a long time since we have had a 'hit'.

Sometimes people will create conflict and chaos around them because they need to get their next 'hit'. They create arguments with people, not because they are trying to make a point or even win, but because they are trying to get a reaction which creates an energy that their emotional body can munch on!

When the shift has been made from horizontal feeding to vertical feeding, a person starts connecting to Source - universal consciousness and unconditional love. They are no longer as interested in feeding off bad news, celebrity gossip or causing pain in others. In fact, these aspects of life will start to become uncomfortable for them. They start moving from a fear-based, mechanistic paradigm to the new authentic space that taps into the universal field.

It is this transition which is being experienced in all aspects of society which will alter the way in which our organizations behave. Whether or not that will happen will depend on what has already been decided in the life plan of the individual. There is no judgment to the shift as better or worse than before - it just is that a lot of people are going through this right now and moving into a New Paradigm.

She Drives Me Crazy

A conscious organization is one that realizes that when there is friction between two people, they are actually playing out emotional issues with one another. They have made a soul agreement to mirror each other and bring these emotions to light. Working with this knowledge does not take the conflict away, but it gives a new level of awareness, so goes a long way to actually resolving it.

Imagine an organizational culture that knows that when issues arise, the participants are co-creating the situation having lovingly decided to hold lessons for each other for maximum soul learning! Instead of just blame and division there is an understanding that conflict is a chance for transformation and growth through the emotional understanding that is gained.

So when conflict arises, the question has to be asked - what are these people reflecting in each other? This question takes the situation from a place of blame to a place of understanding - that there is a gift of learning to be found. Even if the outcome means somebody leaving the company, this is still a perfect outcome of transformation.

The culture that might foster this type of conflict management is not reached overnight. Individuals need to be trained to connect to their own personal process in order to reach this level of understanding. This type of awareness can then become part of the inherent culture of an organization and not just raised in the heat of a conflict situation or it may be seen as gobbledygook as the inner work has not been done to prepare the participants.

Love of the Common People

When our emotional needs are met, when issues that have been carried around since childhood start to be resolved, amazing events can unfold. The vibration of a person literally changes: something that has been validated scientifically in various studies for example, by Professor Valerie Hunt of UCLA.[7] The different vibration attracts a different sort of life. This shift was a process that was already written into that person's life story by their own Higher Self.

Once someone is no longer in the emotional emergency mode, suddenly their world opens up and their sphere of influence and input increases. I witnessed an extremely interesting example of this happening within a group of families who lived in one of the poorest communities in the UK. In 2002, I was called to help out with some corporate training for staff of a government scheme called Sure Start.[8] The lady who called me, called Kim (not the same Kim as I mentioned before), had been working with families out in the East of England who lived in communities receiving a high level of social input.

The British government at the time, had allocated resources for the poorest families in the country with the idea that if children had a better start in life, they would be less of a burden to the state later on. [9] So the scheme targeted the poorest families of the country and the staff were told to be innovative in coming up with solutions to help them. One of their ideas was to contact Kim, a local Reiki Master, who began working with the parents of eleven of the families in a pilot scheme based on empowerment. These were people whom many would see as not being able to change. Some were illiterate and many were out of work. It was extremely innovative of this particular Sure Start scheme to call in Kim as a consultant; the manager of the area was quite a maverick.

Over the period of just a few months Kim worked with the parents and their emotional issues. She also taught them Reiki and they received complementary therapies. It soon became apparent that many of these parents had suffered severe abuse in their lives, as children or even later. They were simply unable to function due to the emotional shut down caused by the abuse. The emotional baggage had made some of them unable to be fully present in their lives. In some, a part of their consciousness had become stuck in another space and time due to the trauma.

By actually resolving some of these emotional issues, they were able to bring themselves back into the present moment. Slowly, they began to train in various skills and also teach others in their neighborhood who had not been on the course, the nature of the material. Some of the parents were glowing with life - as if they had been liberated.

The transformation in the parents and, by association, their children, was so great that Kim was called in again to help train the staff this time. They were so dumbfounded by the shifts in the families that they needed to be trained as to what was going on! For example, some of the teachers didn't understand why children were discussing how to react to bullying with love; it was disturbing to them. Kim called me to provide some of the scientific background to such changes, which is when I was able to meet some of the families and witness some of their stories. Most importantly, when they filled out a survey about their experiences, they regarded their relationships with their children as having greatly improved since the start of the scheme.

Bedtime Stories

It was also liberating for me to witness people being assisted in dealing with their emotional core issues within the public sector. My daily experience as a GP was one that encouraged shoving around emotional issues and never really getting to the core of people's illness or malaise.

In fact Scharmer describes a session with health workers when they uncovered that they all actually believed that illness is a sign of a deeper process happening in a person rather than just the physical, but these deeper aspects are routinely ignored in our Western Health systems.[10] Amazingly, although everybody in the session concurred with this view, they had never spoken about it with each other, nor had done anything to shift their working pattern in order to accommodate their true thoughts, believing that the system was somewhere 'out there' and they could not change it.

This revealed that although many seem to know that we are pushing around the emotional issues in all of our systems - health, education and business, a bit like someone pushes vegetables around a plate if they don't want to eat them, nobody is doing anything about it - our systems simply do not allow for this in their current incarnation e.g. the six-minute GP consultation time. However, it seems that everyone knows that the emotions are best dealt with, just as vegetables are best eaten and will actually feel quite good.

When someone is assisted or starts spontaneously on the path to resolving the emotions that are holding them back, they become more present in their own lives; their sphere of influence changes and less of their consciousness is stuck in the space and time of the trauma - this can even include past lives as discussed previously in this book.

The experience Kim had at Sure Start just illustrates that even people with severe 'problems', as deemed by society, actually respond if core issues are dealt with. These were people who had been relegated to the fringes of society and in some cases had debilitating emotional issues. However, all of us have emotional stories that are playing out in our interactions with the world and with each other. Sometimes they can hold us into the old paradigm philosophies. Sometimes it just may not be our time to move on from the old ways - and that is perfect too.

I Fought the law and the Law Won

Why is it that some people remain stuck in the old ways? It is important to realize that a huge reason why people do not shift is that people are attracted to situations that match their underlying vibration - this is according to the laws of Sympathetic Vibrationary Physics where like vibrations attract and are in resonance with each other.[11] This law also works with our fundamental vibrations of consciousness.

For example, if someone has experienced trauma as a child, they will carry that vibration and will naturally attract situations that match that resonance in their life. So a woman whose father was an alcoholic may end up in a relationship with someone who is addicted to work, because she will naturally resonate with someone who matches this addiction vibration, even though this is not a conscious decision.

Ultimately the behavior of the partner is a wake-up call organized by her self to examine her own patterns, became conscious of them and shift frequency. But until we actually realize this and make the connection between our inner emotional vibration and our seemingly outside world, we will repeat the same patterns and create the same situations. So unless it is in someone's life plan to make the shift, by making the emotional connection, they will stay there until they actually change vibration and therefore resonate at a different frequency of consciousness and thereby attract a different life. If someone has decided in this lifetime that they are *not* going to experience that shift, then that is OK too - they are learning what they, in their great wisdom, have decided to learn.

As we start to understand the dynamics and science of emotions we can better understand the way they are played out within our society and organizations including the powerful effect that the founder has on the whole system via fractal consciousness.

All the Way Down

The person whose emotions matter most to an organization is the founder/CEO or whoever's consciousness is leading the organization (and this may not be so obvious). Their thoughts and emotions will filter from the top to form the basis of the entire organization's culture. This is something that we may sense intuitively, but now our scientific understanding of the cosmos is actually showing us that this is how the universe works - as a fractal.

Cosmology is gradually changing from the Big Bang scenario for our universe, which doesn't fit our data, to incorporating the views of a growing number of scientists who have concluded that the universe is fractal.[12] When you team that with the conclusions of some quantum physicists that consciousness is fundamental to reality have the concept of Fractal Consciousness.

The concept of Fractal Consciousness can be used to give a scientific framework to a phenomenon which until now has been widely assumed, but has not had a scientific basis - that the consciousness of the person in leadership create the prevailing mindset of the whole organization.

All current organizations, therefore, are a microcosm of the whole and displaying fractal behavior. Therefore the consciousness and emotions of whoever is leading the organization and providing the vision and energetic input will be felt throughout the organization. This is something to be aware of - the interior space from where our actions come from.

When it comes to emotions, it is not just our individual shadow that we need to examine, our collective shadow also has an effect as we are all linked. What happens when we make that transition from separation to oneness?

One love, One Heart

The old model is that we are in competition with all life, that the universe is in separation and we have to beat each other by any means necessary due to a belief that the universe is not abundant and we need to claim sparse resources. The new model is that the universe is unified at a deep level: linked through quantum entanglement processes and the central singularity of the black hole which is the infinite light and love - the universal consciousness that is at the center of everything.

So what we see as competition is actually a reflection of us - we are all linked. What we perceive as outside of us is actually reflecting and holding some disowned part of us. This is also true of an organization. Whatever is being projected onto another being is actually a collective disowned emotion.

We saw this all too clearly when the USA was about to invade Iraq. 40% of American people thought that Iraq had been behind the 9/11 bombings.[13] They projected all their disowned hatred onto the Iraqi people and Moslem people in general. They showed their disgust at Saddam Hussain's regime and dictatorship, but then invaded a country that had not attacked them. British and American soldiers killed civilians who were not at war with them.

It was all too easy to project the collective shadow onto Iraq. A projection of a collective shadow tends to occur in any war. Some people may not be aware that wars are deliberately created by the same people who own our monetary systems, but this doesn't negate the fact that people who have no awareness of this plan do actually project their emotional shadow onto whom they perceive as their enemy.

If it is understood that according to the New Science, everything is linked and one, then it is impossible not to take ownership of the shadow. When this shift occurs it becomes clear that what we perceive as the shadow is teaching us something. We then have to integrate that lesson. So our competitors then become our greatest teachers and co-evolvers.

Imagine the shift in our organizations if we were to take the lessons from New Science that the universe is infinitely creative and abundant and that perceived competition is simply a disowned aspect of the collective consciousness which is sending us a message to integrate it again.

In the last two chapters we have explored how the ideas in the New Science discussed in The Genius Groove can relate to various aspects of organizations. It is by no means an exhaustive exploration and this is an area which is set to grow. As we shall see in our last section and as available on The Genius Groove Youtube channel, the ideas of the New Paradigm are already starting to take hold, with leading business visionaries slipping through the cracks of the old world and entering a new dimension.

Part Eight - Talk Talk

Chapter 20

The Genius Groove Interviews

"And we can build this dream together

Standing strong forever

Nothing's gonna stop us now" [u]

Starship

This section contains interviews with people who are exemplify some of the principles people highlighted in The Genius Groove. Please look at The Genius Groove Youtube channel to see videos of these interviews. I feel so blessed for all the people, both featured in the book and on Youtube who have given up their time and of their wisdom to be interviewed for The Genius Groove. Thank you!

Nothing new under the sun: Richard Seymour - Author, journalist and student of Taoism

After having lived the New Paradigm lifestyle for a while, I got a tremendous surprise when I realized that this way of living is nothing new. It was also discovered thousands of years ago by one of the oldest surviving philosophy structures in the world: Taoism.

I met Richard Seymour at a party in Kent in 2003. Within minutes of our meeting he was describing his fascinating novel, *Members Only* which he started working on before he realized he needed to do some serious philosophical study before continuing.[1] You see, the novel takes place in heaven after the hapless Eric finds himself dead. He is told he can only get into heaven if he is a member, but luckily he finds a gold pass in the lavatory, only to find later that it belongs to a gangster who has bribed his way into heaven.

Eric starts conversing with God, which is when Richard needed to take a break to figure out what God was actually going to say. That's how he discovered Taoism. He is now quite an expert and is an active member of major Taoist forums as well as an author of several articles on the subject.

Richard and I shared a lot of ideas whilst writing our respective books. He gave me a copy of a new translation of the *Tao Te Ching* by Wayne L. Wang which merged modern physics with these ancient philosophies.[2]

To my utmost surprise, I found that the lifestyle I call 'New Paradigm Living' is not new at all, but all over the pages of this Taoist text. I interviewed Richard about the principles of Taoism for this book and how they relate to life. He described to me what he has learnt about Taoism from gardening, which can be applied to all of life.

"I looked up how to grow what I wanted to grow. I prepared properly and I bought all the right stuff. I did it all at the correct time of year and I planted it and then I had to sit back and wait. I was quite impatient and every day I went back and checked and nothing was growing and I thought 'God I've done something wrong here'.

And then after a few weeks, the shoots started to appear and I realized that that is part of the essence of Taoism: something called *Wu Wei*, which roughly translates as 'without action'.

A lot of people think it means you don't do anything. But even not doing anything has an effect on what surrounds you anyway, so it's not that simple. What I learnt from that is that you do what you have to do, you act properly, you put in the right preparations and then you have to trust the processes of nature that you've done everything that you can do and then you have to be patient and you cannot force it. You cannot make seeds grow."

Richard gives a great example from Nature about how you cannot push for things to happen, but at the same time you have to be prepared. Please visit the Genius Groove Youtube channel www.youtube.com/geniusgroove for more of the interview and www.richard-seymour.net for more about Richard and his writing.

The Alchemy of Emotions: James Gordon Graham - Entrepreneur, Philosopher and Filmmaker

I can think of no better example of someone who has found that emotional work shifts and aligns your life as my partner, James Gordon Graham. I, along with many others who have heard his story, find it fascinating. In the 1990s he set up a software company which he floated on the stock market eight years later. It is a rare achievement to be CEO from inception to floatation.

Towards the end of his career in the company, he found he could no longer carry on with 'business as usual'. He sought out a different aspect of life and came across teachers of emotional principles. He also made his own conclusions and discoveries about the spiritual and emotional sides of life.

These inner changes, along with many life events caused him to leave the company, sell his shares and start a whole new way of living. Looking at old footage of James on various business TV programs, he is hardly recognizable - such has been his transformation from a hard-edged business man to being fully in touch with his emotions and the spiritual dimensions. It is truly remarkable the speed at which he has learnt what he has.

He is an amazing example of how emotional intelligence can not only transform your company, but it can take it to the next level. His business went to new heights after he got in touch with his own emotions and he, with his staff, created a collective vision for the future. I asked him about the turning point for him and what he thought were the key factors behind his success.

"What I found from a personal point of view was that it was getting harder and harder just chasing money without any kind of values, without any kind of emotional or spiritual angle - it was actually extremely grating... and I was getting more and more tired.

I came to the point where I just realized I don't care what it takes, there must be more to life than this and I reached that point in the mid-90s. It led me on to an understanding of the more emotional, spiritual side of life and especially the importance of working on my emotions. As I became more coherent and more conscious of what I was doing with my emotions, my company grew.

It was very interesting - there was a direct link between my own emotional growth and the growth of the company. The company then developed its products into a more international capability, our profitability grew and also we ended up floating on the stock market in 2001, which very few companies from a pure start up over a period of seven or eight years ever get to. It's a real achievement and it really was due to, as founding

Chief Executive of the company, my ability to become emotionally conscious."

James' new projects are currently under way so keep checking the internet for more details. More snippets of his interview can be found on The Genius Groove Youtube site.

There's always a reason for everything: Rachel Elnaugh - Entrepreneur, author, motivational speaker

Rachel Elnaugh is one of Britain's best known entrepreneurs who shot to fame as the CEO of Red Letter Days, a company with a multimillion pound turnover. As a result she was asked to give advice to budding entrepreneurs in the BBC series *Dragons Den*.[3] When her company went into administration, she took all the lessons she learnt and put them into her book *Business Nightmares*.[4]

She is now a popular motivational speaker with a new company that teaches a deeper wisdom then just profit. I was introduced to Rachel by a mutual friend and what struck me is her wisdom, which is not always given a chance to breathe in her public image. Her interests are wide ranging, from politics to the Law of Attraction. She is able to weave all this together along with her tremendous experience at the helm of one of Britain's most prominent companies into a very interesting philosophical package. When I went to interview her, what I was not expecting was for her to be Living New Paradigm too! She spoke to me about how living in this way actually improves her business.

"In my Red Letter Days era, I very much pushed to make stuff happen in a very alpha way using sheer force and post that, I've actually worked in a much gentler, more energetic way, going with energy and opportunity as opposed to trying too much from left brain logic. And actually it's far more powerful and far more exciting.

When you're using your left brain you're controlling everything and it's all about you and your ego making stuff happen the whole time and you get ever decreasing circles actually.

Whereas when you start working from the right brain and working with your energy, you start just being really open and saying 'yes'. I'm really open and receptive now to opportunities and seeing where it takes me."

Please visit The Genius Groove Youtube site for more snippets and also Rachel Elnaugh's website www.rachelelnaugh.com for details of her courses and her book.

Geniiosity! Soleira Green - author, visionary, business speaker

If there is one person whose eloquence makes my jaw drop, it is Soleira Green. We worked together previously when she was a speaker at the Children of the New World conferences and I caught a glimpse of just how far she and some of her friends are actually taking consciousness. This is not a person who follows trends or even makes them. She actually consciously creates from beyond the edge of perception and alters the dynamics of our dimension.

She had some amazing ideas on genius and creativity which really summarize the new type of genius.

"There was a time in my life when I wouldn't have considered myself a genius and I now do consider myself that and I believe that everybody has the capacity to be a genius and I don't think I mean genius as in a measurement of IQ or intellectual quotient.

Genius gets expressed in many ways and it seems to come both from deep inside people. There's something that's uniquely their gift, their contribution that's extraordinary, that's really gifted in their own right. But it also comes from outside them. So there's a two way flow happening here with genius.

To me people like Da Vinci and Mozart and Beethoven, people that we would consider geniuses, I think they absorbed their art from the air. I think the inspiration for it came from within them and there's a link that marries from the external genius pool, if you will, all the things that are waiting to be brought into being and the internal genius pool which is where your unique essence and your unique passion drives from.

And there's something where the two meet that creates realized genius in the world. And I believe that that act of meeting is something that can be learned and trained up. I don't think that genius is something that a few people are born with and the rest of us poor suckers don't get any! That's just not an accurate representation."

For more information on Soleira Green, please visit one of her website www.soleiragreen.com. There is more from this interview on The Genius Groove Youtube site.

The Work we were Born to Do: Nick Williams - Leading work expert, author, business speaker

Like many people, I heard about Nick through his work with Alternatives in St James' Church Piccadilly, London and from his numerous lectures at Mind Body Spirit events. But it was not until I started writing this book, that I picked up *his* classic book, *The Work we were Born to do* and was astounded by its gentle depth which is much like the man himself. [5] Nick started out in the IT industry and did well in that, but found his life lacked meaning which is when he was drawn to become a work expert and help others on their journey.

When he spoke to me, he imparted some really good advice on how to follow your passion and Genius Groove, but also make money as well as discussing how a lot of emotions come up when we move towards our goals.

"Two beliefs that a lot of us have grown up with is that to make a lot of money you have to sell your soul, you have to sell out your deeper self and just show up and do a job and sell out on your own values and beliefs to some extent. 'Just get over it, that's what you have to do, life is about suffering'.

And then the other belief is that, OK if you want to defy that you can go off and follow your heart and do something inspiring that you're really passionate about, that's really meaningful to you, but give up any hope of earning much money or having any financial security.

So it's almost like the love or the money, you can't have both. That's such a deep belief in most people's psyches that they can't have both... For some people I think it's this belief in sacrifice so you either sacrifice the money for the love or the love for the money and I think it takes a lot of courage to defy that and say actually I'm going to have both because for many people it hits so much guilt and so much unworthiness that's hidden away. I think to have a really blessed life where you're happy, where you're doing what you love, you have financial abundance and everything is going great - most of us think we are going to get so much jealousy and attack and criticism come our way if we have a really blessed life.

In my experience you have to be entrepreneurial about living your purpose... Find your work, put it out in the world and get clever about getting paid for doing it. Most people are thinking, how can I get paid for this before I start doing it. In my experience, you just have to go out there and start doing it. Show up in the world do what you love and get clever about how do I find the people that want it and how will they pay me for it. Ultimately, it's not about the money. We're spiritual beings and it's about putting our heart and spirit in the world, creating enough money and we can live the life that we want."

For the rest of the interview, please visit the Youtube channel and for more about Nick Williams visit, www.inspired-entrepreneur.com.

Follow the Calling: Molly Harvey - Inspirational Speaker, author and The Corporate Soul Woman.

There's something about Molly Harvey that makes you feel at peace. It seems that I am not the only one who feels this, as many large corporations have booked her as a speaker to bring some of her Irish charm their way. For me, she is a soul sister and a genuine mystic who is just as at home amongst CEOs as she is in a meditation meeting. She has a strong thirst for information and her home is littered with the latest inspirational business books as well as her favorites.

Her accolades are many including being the first woman president of the Professional Speaker's Association. Although she has spent many years in a corporate environment, she is standing true to who she is by speaking about the soul. In her interview she was incredible as always, but one of the aspects I would like to highlight is when she spoke about our inner calling.

"I'd say to anybody out there right now, if they are at a cross roads in their life and there's a deeper sense of calling. I would say, follow the calling. Follow it. Because everyday you follow it, it's a funny thing, it's almost like you get a deeper sense of courage. And you'll get past the dragons at the gate. and don't be thinking there won't be many more dragons, because there will and the tests get more challenging, but as they do in life, you'll find a deeper sense of purpose in your life, whatever purpose is for you."

To find out more about Molly, please visit
www.thesoulwoman.com

Strike A Pose - There's Nothing To It

And that just about wraps things up for our journey in The Genius Groove. I am going to leave you now to follow The Call to a more authentic life, open up to New Paradigm Living and get into *your* true Genius Groove! To quote once more, from Madonna, from her Desperately Seeking Susan era,

"C'mon,

I'm Waiting!"

Afterword

I look out to the mists over Lake Toya, not quite believing I'm here. Writing that book in my little office has led to *this*?

I look around at my huge Japanese hotel suite. There's not a lot of time to take it all in. We have a busy schedule. I have to join Brian and the others for the rest of the symposium.

We are talking about my favorite topic - science, consciousness and spirituality.

All trials, tribulations and triumphs have led to this place. This and every moment, is where I find my bliss.

This is my Genius Groove.

Guide to Abbreviations

BBC - British Broadcasting Corporation
BHP - Black Hole Principle
CEO - Chief Executive Officer
CIA - Central Intelligence Agency
CV - Curriculum Vitae
DKNY - Donna Karen New York
DNA - Deoxyribonucleic Acid
EFT- Emotional Freedom Technique
GMC - General Medical Council
GP- General Practitioner
IBM - International Business Machines Corporation
IQ - Intelligence Quotient
IT - Information Technology
LOA- Law of Attraction
MIT - Massachusetts Institute of Technology
NLP- Neuro Linguistic Programming
PO Box - Post Office Box
(Micro) Quasar - (Micro) Quasi-Stellar Radio Source
TV - Television
UCLA - University of California Los Angeles
UK - United Kingdom
VSL - Variable Speed of Light

<u>Suggested Reading/Viewing</u>

<u>Books</u>

The Work We were Born to Do - Nick Williams
Healing is a Daily Business - Lucia Nella
The Dark Side of the Light Chasers - Debbie Ford
The Breakthrough Experience - Dr John Demartini
Faster than the Speed of Light - Joao Magueijo
The Biology of Belief - Bruce Lipton
Members Only - Richard Seymour
The Element - Sir Ken Robinson
The Field - Lynne Mc Taggart
Science and the Akashic Field - Ervin Laslo
Punk Science - Manjir Samanta-Laughton
The Conscious Universe - Amit Goswami
From Science to God - Peter Russel
Business Nightmares - Rachel Elnaugh
Born on a Blue Day - Daniel Tammet
Women who run with Wolves - Clarissa Pinkola Estes
Eat, Pray, Love - Elizabeth Gilbert
The Alchemical Coach - Soleira Green

<u>Films available on the internet</u>

Zeitgeist Addendum
JK Rowling's Harvard Commencement Speech
Steve Jobs' Stanford Commencement Speech
Sir Ken Robinson's TED lecture www.TED.com
Elizabeth Gilbert's TED Lecture www.TED.com
www.youtube.com/geniusgroove

DVDs

The Secret - Rhonda Byrne
What the Bleep do we know? - Vincente M, Chasse B, Arntz W.
Canticle to the Cosmos - Brian Swimme
Personifying the Quantum Theory - Dr John Demartini
The Living Matrix - Greg Becker, Harry Massey.
The Money Masters: How International Bankers Gained Control of America. Carmack P.
Money As Debt. Grignon P.

References

Introduction

1. Samanta-Laughton M. *Punk Science*. (O-books) 2006.
2. Dawkins R (presenter) *The Enemies of Reason*. [DVD] (Siren Visual) 2008.

Chapter 1 - Groove Control

1. White M. *Isaac Newton:The Last Sorcerer*. (Fourth Estate) 1998.
2. 2007 *Annual Survey of Hours and Earnings First Release*. Department of National Statistics. 2007
http://www.statistics.gov.uk/StatBase/Product.asp?vlnk=15050 [cited August 2009].
3. Snow D. Snow J. (presenters) *What Britain Earns*. BBC2. Aired January 2008.
4. The Best (and worst) paid jobs in Britain. *Daily Mail*. 10th January 2008.
5. Ibid.
6. Balakrishnan A. UK's decade of House Price Growth. *The Guardian*. 2nd February 2008.

Chapter 2 - Spirits in the Material World

1. Icke D. (subject) *The Terry Wogan Show*. BBC 1. Aired 1991.
2. Rowbotham M. *The Grip of Death*. (Jon Carpenter Publishing) 1998.
3. Ibid.
4. Ibid.
5. Grignon P. *Money As Debt*. [DVD] Moonfire Studio.

6. Carmack P. *The Money Masters: How International Bankers Gained Control of America.* (Royalty Production Company) [DVD] 1998.
7. http://www.federalreserve.gov/ [cited January 2008].
8. http://en.wikipedia.org/wiki/Bank_of_England [cited June 2008].
9. Cocking M. *Essay on the Royal Charter for Bank of England.* June 2009. (Available on request from author).
10. Carmack P. *The Money Masters: How International Bankers Gained Control of America* (Royalty Production Company) [DVD] 1998.
11. *Brown Loses Grip on Bank.* BBC Politics 97. http://www.bbc.co.uk/politics97/news/05/0506/brown.shtml [cited August 2009].

Chapter 3 - Karma Chameleon

1. Miller G. *The Mating Mind.* (Vintage) 2001.
2. Robinson K. *The Element.* (Allen Lane) 2009.
3. Ibid.
4. Ibid.
5. Lipton BH. *The Biology of Belief.* (Mountain of Love) 2005.
6. Orwant R. What makes us human? *New Scientist.* 21 February 2004; 36-39.
7. Myers, EW. et al. A whole-genome assembly of Drosophila. *Science.* 24, March 2000; 287: 2196-2204.
8. Goff SA, et al. A draft sequence of the rice genome (Oryza sativa L. ssp japonica) *Science* 5 April 2002; 296: 92-100.
9. Orwant R. What makes us human? *New Scientist.* 21 February 2004; 36-39.
10. Ibid.

11. Carroll, Sean B. et al. Regulating Evolution. *Scientific American*. May 2008;60–67.
12. Lipton BH. *The Biology of Belief*. (Mountain of Love) 2005.
13. Ibid.
14. Ibid.
15. Robinson K. *Out of Our Minds*. (Capstone) 2001.

Chapter 4 - Revolution Baby

1. Kurzweil R. *The Singularity is Near*. (Penguin). 2006.
2. http://bluebrain.epfl.ch/ [cited August 2009].
3. Greenfield S. *The Brain Consciousness and Controversies*. Lecture given to the Royal Institution, 10th November 2008.
4. Chalmers D. *The Conscious Mind*. (Oxford University Press) 1996.
5. Fields RD. Making Memories Stick. *Scientific American*. February 2005; 58-65.
6. Ibid.
7. Pribram, KH. The Neurophysiology of Remembering. *Scientific American*. January 1969; 73-86.
8. Vincente M, Arntz W, Chasse B. What the Bleep do we know? [DVD]. (Revolver Entertainment) 2005.
9. Krips H. Measurement in Quantum Theory. *The Stanford Encyclopedia of Philosophy*. Winter 1999 Edition. http://plato.stan-ford.edu/archives/win1999/entries/qt-measurement. [cited January 2006].
10. Goswami A. *The Self-Aware Universe*. (Tarcher/Putnam) 1995.
11. Russell P. *From Science to God: Exploring the Mystery of Consciousness*. (New World Library) 2005.
12 Goswami A. *The Self-Aware Universe*. (Tarcher/Putnam) 1995.

13. Adams D. *The Hitchhiker's Guide to the Galaxy*. (Pan MacMillan) 1979.
14. Goswami A. *The Self-Aware Universe*. (Tarcher/Putnam) 1995.
15. Ibid.

Chapter 5 - The New User's Guide to the Brain

1. Parker S. (subject) *The Woman With No Brain*. Extraordinary People Series. (ZKK Productions) 2003.
2. Bilimoria E. *Consciousness came first: Connecting non-locally with the whole*. Presentation at conference on Esoteric Perspectives on a science of consciousness. [PDF file] Scientific and Medical network. 31 May 2003, Rudolph Steiner House, London.
3. Holt J. *Blindsight and the Nature of Consciousness*. (Broadview Press) 2003.
4. Dennett D.C. *Consciousness Explained* (Back Bay Books) 1992.
5. Penrose R. *The Emperor's New Mind*. (Oxford University Press) 1989.
6. Bortman H. Energy Unlimited. *New Scientist*. 22 January 2000; 32- 34.
7. Lucas G. (Director) *Star Wars*. [DVD] (20th Century Fox) 2008.
8. http://www.musion.co.uk/ [cited August 2009]
9. Chown M. All the world's a hologram. *New Scientist*. 17 January 2009.
10. Bekenstein J.D. Information in the Holographic Universe. *Scientific American*. August 2003; 48-55.
11. Pribram, KH. The Neurophysiology of Remembering. *Scientific American*. January 1969; 73-86.

12. Boly M et al. Perception of pain in the minimally conscious state with PET activation: an observational study. *The Lancet Neurology*. Nov 2008: Vol. 7 No. 11 pp 1013-1020.

13. Sabom M. *Light and Death*. (Zondervan) 1998.

14. Lagerfeld K. (subject) *The Secret World of Haute Couture*. (BBC) 2006.

15. Laszlo E. *The Whispering Pond: a Personal Guide to the Emerging Vision of Science*. (Element) 1999.

16. Talbot M. *The Holographic Universe*. (Harper Collins) 1996.

17. McTaggart L. *The Field*. (Harper Collins) 2001.

18. Schwartz GE, Simon WL. *The Afterlife Experiments*. (Simon and Schuster.) 2002.

19. Greene Brian. *The Elegant Universe*. (Random House) 2000.

20. Mandeville M. *Cayce and the Economy*. Phoenix Book Two. (Published online) http://www.michaelmandeville.com/phoenix/trilogy/booktwo/rp2chapter23.htm [cited August 2009].

21. Pond D. ed. *Universal Laws Never Before Revealed*. (Infotainment world Books) 1995.

22. Kaku M. *Hyperspace*. (Oxford Paperbacks) 1995.

23. Holt J. *Blindsight and the Nature of Consciousness*. (Broadview Press) 2003.

24. Tammat D. (Subject) *The Boy with the Incredible Brain*. Extraordinary People Series. Focus Productions. 2005.

25. Tammat D. *Born on a Blue Day*. Hodder & Stoughton Ltd 2006.

26. Abraham C. *Possessing Genius: the Bizarre Odyssey of Einstein's Brain*. (Icon) 2005.

27. Witelson SF, Kigar DL, Harvey T, The Exceptional brain of Albert Einstein. *Lancet*. June 1999; 353(9170): 2149-53.

28. Fields D. R. Stevens-Graham B. New Insights into Neuron-Glia Communication. *Science*. October 18 2002; 298: 556-562.
29. Maguire,E.A., Woollett,K., Spiers,H.J. London taxi drivers and bus drivers: A structural MRI and neuropsychological analysis. *Hippocampus*. 2006;16(12): 1091-1101.
30. Yang Y et al. Prefrontal white matter in pathological liars.*The British Journal of Psychiatry* 2005;187: 320-325.
31. Perry, BD and Pollard, D. Altered brain development following global neglect in early childhood. *Society For Neuroscience: Proceedings from Annual Meeting*, New Orleans, 1997.

Chapter 7 - Borderline

1. Demartini J F. *The Breakthrough Experience*. (Hay House) 2002.
2. The International Conferences on Science and Consciousness. The Message Company. www.bizspirit.com [cited August 2009].
3. Swimme B. *Canticle to the Cosmos*. [DVD/VHS] Tides Center. 1990.

Chapter 8 - Waiting for a Star to Fall

1. Perlmutter S et al. Discovery of a supernova explosion at half the age of the universe and its cosmological implications. *Nature*.1 January 1988; 391: 51-54.
2. Magueijo J. Faster than the speed of *light*. (Arrow) 2004.
3. Hawking S. Penrose R. *The Nature of Space and Time*. (Princeton University Press) 2000.
4. Hawking S. *A Brief History of Time*. (Bantam) 1988.
5. Ibid.
6. Ibid.

7. Henbest N. The Great Annihilators. *New Scientist*. 1 April 2000; 28-31.

8. Ibid.

9. Swimme B. *Canticle to the Cosmos*. [DVD/VHS] Tides Center. 1990.

10. Mandlebrot B. Fractals and Chaos: The Mandelbrot Set and Beyond. (Springer) 2004.

11. Gehrels N, Piro L, Leonard PJT. The Brightest explosions in the Universe. *Scientific American*. December 2002; 53-59.

12. Amelino-Camelia G. Double Special Relativity. *Nature*. 2002; 418: 34-35. http://arxiv.org/abs/gr-qc/0207049.

13. Gehrels N, Piro L, Leonard PJT. The Brightest explosions in the Universe. *Scientific American*. December 2002; 53-59.

14. Schilling G. Do black holes play with their food? *Science NOW*. 18 August 2005; 4.

15. Finkbeiner DP. *WMAP Microwave Emission Interpreted as Dark Matter Annihilation in the Inner Galaxy*. January 2005. http://www.arxiv.org/abs/astro-ph/0409027.

Chapter 9 - Velocity Girl

1. Brooks M. The Results are in... and now it's time to party. *New Scientist*. 5 April 2003; 22-23.

2. Filippini J The Standard Cosmology http://cosmology.berkeley.edu/Education/CosmologyEssays/The_Standard_Cosmology.html [cited August 2009].

3. Magueijo J. *Faster than the speed of light*. (Arrow) 2004.

4. William R. *Speed of Light* (abc.science) 2000. http://www.abc.net.au/rn/science/ss/stories/s212674.htm [cited Dec 2005].

5. Milewski JV. Superlight, *One Source one force*. 17 November 1996.http://www.luminet.net/~wenonah/new/milewski.htm. [cited December 2005].

6. Smith DM, Lopez LI, Lin RP, Barrington-Leigh CM. Terrestrial Gamma-ray flashes observed up to 20 MeV. *Science*. 18 February 2005; 307: 5712:1085-1088.

7. Richter P, Wakker BP. Our growing breathing galaxy. *Scientific American*. January 2004; 28-37.

8. Gefter A. Don't Mention the 'F' word. *New Scientist*. 10 March 2007; 30-33.

9. Hunt V. Infinite Mind. (SOS Free Stock) 1996.

10. Cartlidge E. Half the Universe is missing. *New Scientist*. 4 September 2004; 37-39.

11. Tiller WA. *Science and human transformation*. (Pavior) 1997.

12. Larson DB. *Nothing but Motion*. (North Pacific Publishers) 1979.

13 Gefter A. The world turned inside out. *New Scientist*. 20 March 2004; 34-37.

14. Green M. The Synthesis of relativity and Quantum theory in String theory. 11 April 2005. Plenary presentation at the Institute of Physics: a century after Einstein, University of Warwick.

15. Chown M. Shadow Worlds. *New Scientist*. 17 June 2000; 36-39.

16. Lerner E. Bucking the Big Bang. *New Scientist*. 22 May 2004; 20.

17. Fox M. *Creativity: where the divine and human meet*. (Tarcher) 2004.

Chapter 10 - What a Feeling!

1. Denempont B. Original Email announcing Eli scandal.
http://home.comcast.net/~sresnick2/gangaji_emails.htm
[cited August 2009].
2. Jaxon-Bear E. *Taking Full Responsibility*. Recording of talk
Australia 2008.
http://www.youtube.com/watch?v=zjoiSv0lvSE [cited August
2009].
3. Gilbert E. *Eat, Pray, Love*. (Bloomsbury) 2007.

Chapter 11 - Second that Emotion

1. Ford D. *The Dark Side of the Light Chasers*. (Mobius) 2001.
2. Myss C M, Shealy N C. *The Creation of Health*. (Bantam) 1999.
3. Nella L. *Healing is a Daily Business*. (Lightstar publishing)
2008.
4. Demartini J F. *Personifying the Quantum Theory*. [DVD] (The
Demartini Human Research and Education Foundation) 2002.
5. Demartini J F. *The Breakthrough Experience*. (Hay House) 2002.
7. Ford D. *The Dark Side of the Light Chasers*. (Mobius) 2001.
8. Ciconne M, Orbit W. *Mer Girl*. (Maverick) 1998.
9. Mother Teresa. Come be my Light. (Rider & Co) 2008.
10. Dyson M E. *I may not get there with you*. (Free Press) 2001.
11. Ford D. *Why Good people do Bad things*. (HarperOne) 2009.
12. Ford D. Amazon.com blog
http://www.amazon.com/Debbie-Ford/e/B001I9Q7G8/ref=n
tt_dp_epwbk_0 [cited August 2009.

Chapter 12 - The Time of my Life

1. Tolle E. *The Power of Now*. (New World Library) 2004.
2. Tolle E. A Happier You. *O Magazine*. January 2009;129.

Chapter 13 - Same as it Ever was

1. Brooks M. The Weirdest Link. *New Scientist*. 27 March 2004; 32-35.
2. Gefter A. The world turned inside out. *New Scientist*. 20 March 2004; 34-37.
3.Larson DB. *Nothing but Motion*. (North Pacific Publishers) 1979.
4. Leibovici L. Effects of remote, retroactive intercessory prayer on outcomes in patients with bloodstream infection: random-ised controlled trial. *BMJ* 2001;323:1450-1451.
5. Puthoff H, Targ R. *Mind-reach*. (Hampton Roads) 2005.
6. Ibid.
7. Sugrue T. The Story of Edgar Cayce. (ARE Press) 1997.
8. Mandeville M. *Cayce and the Economy*. Pheonix Book Two. (Published online.) http://www.michaelmandeville.com/phoenix/trilogy/booktwo/rp2chapter23.htm [cited August 2009].
9. Jaworski J. *Synchronicity: the inner path of leadership*. (Berrett-Koehler) 1998.
10. Spears B. (subject) *The Rise and Rise of Britney Spears*. VH1 documentary.
11. Gladwell M. *Outliers*. (Allen Lane) 2008.

Chapter 14 - Please, Please, Please let me get what I want

1. Byrne R. *The Secret*. (Simon and Schuster) 2006.
2. Byrne R. *The Secret*. [DVD] (Prime Time Productions). 2006.
3. *My Kid's Psychic*. World of Wonder Documentary for Channel Four. 2006.
4. Lucas G. (Director) *Star Wars*. [DVD] (20th Century Fox) 2008.
5. Water S. *You Are God, Get Over It!* (Limitlessness) 2005.
6. Winfrey O. (presenter) *Spirituality 101 Day Three*. (Harpo Productions.) Broadcast 2008.

Chapter 15 - I Want to Break Free

1. Mathur P. *The Sickening Mind*. Harper Collins. 1997.
2. Dossey L. *Space, Time and Medicine*. (Shambala) 1982.
3. Cruttenden W. *Lost star of Myth and Time*. (St Lynn's Press) 2005.

Chapter 16 - Groove is in the Heart

1. Pinkola Estes C. *Women who run with the Wolves*. (Rider and Co) 1993.
2. Alexander, B.K., Beyerstein, B.L., Hadaway, P.F., and Coambs, R.B. Effect of early and later colony housing on oral ingestion of morphine in rats. *Pharmacology Biochemistry and Behavior*. 1981; Vol 15, 4:571-576.
3. Gladwell M. *Outliers*. (Allen Lane) 2008.
4. Davies G. (subject) Per Una launch documentary. British Television. 2001.
5. Lin D. Lao Tzu. *Tao Te Ching*. (Skylight Paths) 2007.
6. Murray W H. *The Scottish Himalayan Expedition*. (Dent) 1951.

7. McTaggart L. *The Field*. (Harper Collins) 2001.

Chapter 17 - Shaking the Tree

1. Sell C. www.heaven-on-earth.co.uk. [cited September 2009].
2. Gilbert E. *Eat, Pray, Love*. (Bloomsbury) 2007.
3. Sellers P. The Business of Being Oprah. *Fortune*. 1 April 2002; 62-69.
4. Ramsey G.(subject) *Friday Night with Jonathan Ross*. (BBC1) 2004.
5. Water S. *You Are God, Get Over It!* (Limitlessness) 2005.
6. Ford D. *Why Good people do Bad things*. (HarperOne) 2009.
7. Ford D. *The Dark Side of the Light Chasers*. (Mobius) 2001.

Chapter 18 - New Power Generation

1. Barrett R. *Liberating the Corporate Soul*. (Butterworth-Heinemann) 1998.
2. Wheatley MJ. *Leadership and the New Science*. (Berrett-Koehler) 1999.
3. Guillory W A. *The Living Organization*. (Innovations International) 1997.
4. Barrett R. *Liberating the Corporate Soul*. (Butterworth-Heinemann) 1998.
5. Scharmer C O. *Theory U:Learning from the Future as it Emerges*. (Berrett-Koehler) 1999.
6. Senge P, Scharmer C O, Jaworski J. Flowers B S. *Presence: Human Purpose and the Field of the Future*. (Currency) 2008.
7. Semler R. *Maverick*. (Random House) 2001.
8. Scharmer C O. *Theory U:Learning from the Future as it Emerges*. (Berrett-Koehler) 1999.

9. www.changetheworldcorp.com [cited September 2009].

10. The Sunday Times Bestsellers week ending 25/07/09. *The Sunday Times*. 2nd August 2009; 32.

11. Byrne R. *The Secret*. (Simon and Schuster) 2006.

12. Wheatley MJ. *Leadership and the New Science*. (Berrett-Koehler) 1999.

13. Capra F. *The Tao of Physics*. (Flamingo) 1992.

14. Capra F. *The Turning Point*. (Flamingo) 1983.

15. Barrett R. *Liberating the Corporate Soul*. (Butterworth-Heinemann) 1998.

16. Semler R. *Maverick*. (Random House) 2001.

17. Hickman M. Streets ahead: Does John Lewis offer a revolutionary way forward for big business? *The Independent*. Thursday, 20 August 2009.

18. Introducing the Brand New Social Enterprise Awards. http://www.socialenterpriseawards.org.uk/pages/socialenter priseawards.html [cited September 2009].

19. Evans G W. *Environmental Stress*. (Cambridge University Press) 1984.

20 Samanta-Laughton M. *Punk Science*. (O-books) 2006.

21. Tiller W. A. *Conscious Acts of Creation*. (Pavior) 2001.

22. Scharmer C O. *Theory U:Learning from the Future as it Emerges*. (Berrett-Koehler) 1999.

23. Jaworski J. *Synchronicity: the inner path of leadership*. (Berrett-Koehler) 1998.

24. Scharmer C O. *Theory U:Learning from the Future as it Emerges*. (Berrett-Koehler) 1999.

25. Tiller WA. *Science and human transformation*. (Pavior) 1997.

26. Ibid.

27. Wheatley MJ. *Leadership and the New Science*. (Berrett-Koehler) 1999.

28. Hamer RG. *Summary of the New Medicine.* (Amici de Dirk) 2000.

29. Scharmer C O. *Theory U:Learning from the Future as it Emerges.* (Berrett-Koehler) 1999.

30. Ibid.

31. Wheatley MJ. *Leadership and the New Science.* (Berrett-Koehler) 1999.

32. Semler R. *Maverick.* (Random House) 2001.

33. Caulkin S. Who's in Charge here? *The Observer.* 27 April 2003.

Chapter 19 - My Ever Changing Moods

1. Cruttenden W. *Lost Star of Myth and Time.* (St. Lynn's Press) 2005.

2. Ibid.

3. Scharmer C O. *Theory U:Learning from the Future as it Emerges.* (Berrett-Koehler) 1999.

4. Ibid.

5. Barrett R. *Liberating the Corporate Soul.* (Butterworth-Heinemann) 1998.

6. Nella L. *Healing is a Daily Business.* (Lightstar publishing) 2008.

7. Hunt V. Electronic Evidence of Auras and Chakras in UCLA study. *Brain/Mind Bulletin.* 1978; 3:9.

8. *Meeting the Childcare Challenge.* Green Paper. (SureStart Publications.) 2004.

9. Ibid.

10. Senge P, Scharmer C O, Jaworski J. Flowers B S. *Presence: Human Purpose and the Field of the Future.* (Currency) 2008.

11.Pond D ed. *Universal Laws Never Before Revealed*. (Infotainment world Books) 1995.

12. Gefter A. Don't Mention the 'F' word. *New Scientist*. 10 March 2007; 30-33.

13. Scharmer C O. *Theory U:Learning from the Future as it Emerges*. (Berrett-Koehler) 1999.

Chapter 20 - The Genius Groove Interviews

1. Seymour R. *Members Only*. (Exposure) 2006.

2. Wayne W.L. Dynamic Tao and it's manifestations. (Helena Island Publisher) 2004.

3. Elnaugh R. (subject) *Dragon's Den* Series 1/2. BBC 2. 2005.

4. Elnaugh R. *Business Nightmares*. (Crimson) 2008.

5. Williams N. *The Work Were Born to Do*. (Element) 2000.

Musical References

a) Michael McDonald. *Sweet Freedom* by Rod Temperton. (Warner).

b) Madonna. *Spotlight* by Madonna and Stephen Bray. (Sire).

c) Police. *Spirits in a Material World* by Sting. (A&M).

d) Edie Brickell and the New Bohemians. *What I Am* by Edie Brickell. (Geffen).

e) Transvision Vamp. *Revolution Baby* by N C Sayer. (MCA).

f) Big Audio Dynamite. *C'mon every Beatbox* by M Jones, D Letts. (CBS).

g) Ziggy Marley and the Melody Makers. *Black My Story* by Z Marley, S Marley. (Virgin).

h) Madonna. *Over and Over* by Madonna and Stephen Bray. (Sire).

i) Steve Winwood. *Higher Love* by Steve Winwood, Will Jennings. (Island).

j) Madonna. *Ray of Light* by William Orbit, Clive Muldoon, Dave Curtiss and Christine Leach. (Maverick).

k) Charlene. *I've never been to me*. Ron Miller, Kenneth Hirsch. (Motown).

l) Madonna. *Nothing really matters* by Madonna, Patrick Leonard. (Maverick).

m) Talking Heads. *Road to Nowhere* by David Byrne, Tina Wymouth, Chris Frantz, Jerry Harrison. (Warner/Chappel).

n) Steve Winwood. *Valerie* by Winwood, Jennings. (Island).

o) Frankie Goes to Holllywood. *Relax* by Gill, Johnson, Nash, O'Toole. (ZTT).

p) Madonna. *Substitute for Love* by Madonna, William Orbit, Rod McKuen, Anita Kerr, David Collins. (Maverick).

q) Chesney Hawkes. *The One an Only* by Nik Kershaw. (Chrysalis).

r) Huey Lewis and the News. The Power of Love by C. Hayes, H. Lewis, J. Colla. (Chrysalis).

s) Paul Mc Cartney and the Frog Chorus. *We all stand Together* written by Paul Mc Cartney. (Parlophone).

t) Madonna. *Bedtime Story* by Björk , Marius DeVries, Nellee Hooper. (Maverick).

u) Starship. *Nothing's gonna stop us now* by Albert Hammond and Diane Warren. (RCA/BMG).

About the Author

Dr Manjir Samanta-Laughton is a medical GP turned international author and lecturer. She highlights the links between cutting-edge science and spirituality. In 2006, her first book, *Punk Science* was published by O-books. It became an international sensation, revealing a new vision of the cosmos that places black holes at the creative heart of the universe. Her second book, *The Genius Groove* is published by Paradigm Revolution Publishing.

Manjir is a popular speaker and has toured in the UK, USA and Europe. She has done many interviews for TV, documentaries, radio as well as national press including the BBC and C4. She lives in Derbyshire with her partner, James.

To find out more about Manjir and her work and to sign up for exclusive content please visit the following websites

www.thegeniusgroove.com
www.punkscience.com
www.paradigmrevolution.com

Your Life just Got Groovy!

Lightning Source UK Ltd.
Milton Keynes UK
17 November 2009

146365UK00002B/110/P